NICKEL COUNTRY
- GOLD COUNTRY

NICKEL COUNTRY - GOLD COUNTRY

Norma King

RIGBY

Dedicated to my husband
and children

RIGBY LIMITED • ADELAIDE • SYDNEY
MELBOURNE • BRISBANE • PERTH

First published 1972
Copyright © 1972 Norma King
Library of Congress Catalog Card Number 70-160566
National Library of Australia Card Number
and ISBN 0 85179 389 4

Printed in Hong Kong

CONTENTS

ILLUSTRATIONS

ACKNOWLEDGMENTS

Thanks are given to West Australian Newspapers Ltd, the *Kalgoorlie Miner*, the *Independent*, and the *Sunday Times* for allowing me the use of valuable material from their publications; to Western Mining Corporation and Bob Gude for photographs and use of material from the *Westminer*; to Ted Mayman for his encouragement and advice; to John K. Ewers for trying to cure my "passive verbs;" to Maureen Timoney for her continued interest; and to Mr R. Mills, President of the Eastern Goldfields' branch of the Historical Society for the use of his scrapbook and material from the Society; to J. A. Mitchell and Mark Randell and those other people who loaned me photographs; to Mr A. Pritchard in Norseman, and to Mr Schoord for material from his thesis on Norseman; to the various Mining and Exploration Companies for information; and for the help given by Bill Moriarty, Fred Smith, Lindsay Burrows, Bernie Rutland, and to all those mentioned in the book who told me their stories. Acknowledgements are made for material from *Those Were the Days* by Arthur Reid.

Special thanks are given to Mrs J. O'Brien for taking copies of historical photographs and for the loan of some of her own slides, and last, but not least, to the various members of my family for assistance given in different ways and for putting up with me while the book was being written.

INTRODUCTION

The nickel boom made very little sound at first. In mining circles the noise was louder, but to the man in the street it was only a low rumble, a thing to excite interest for a while then fade away. Nickel seemed a rather poor substitute after Gold had been King for so long.

Then Kambalda began operating, and other companies started their exploration work, at first close to Kalgoorlie, then spreading out in ever widening circles, and the sound of the boom got louder and louder till, after the finding of nickel by Poseidon, its thunderous roar reverberated throughout the world.

1
BEFORE NICKEL

THE TWIN TOWNS of Kalgoorlie and Boulder in 1965 B.N. (Before Nickel) were rather depressing places to live in. There were voices of gloom everywhere. We were told that the mines could not last much longer, and that a rise in the price of gold was the only thing to save us.

They were almost becoming old folk's towns. We were having difficulty in keeping our children here. After leaving school, all they wanted to do was to "get out of this dump."

Apprenticeships, nursing training, or courses at the Technical School or School of Mines kept some here longer. But when their training was finished, there was no holding the youngsters.

To go north was the thing to do. Things were booming there, with the new iron-ore mines, their ports, and ample opportunities for all. They were magnets that drew a lot of our men, young and not so young, who told us that there was no future in Kalgoorlie and Boulder.

The general uneasiness and lack of confidence were reflected in the appearance of the towns. They looked like rather shabby old ladies who had known better days, as indeed they had. People thought twice before painting their homes or replacing leaning, rusty fences. The many families leaving the towns sold their homes cheaply.

Civic minded citizens had been talking about secondary industries for some time, but the distance of 375 miles from Perth, and high freight charges, soon ruled these out.

Various ideas were put forward to keep the towns going, and one was a suggestion that we turn Kalgoorlie into a second Las Vegas. We already had a tarnished image with our "Two-up School," extended drinking hours, and the professional "ladies" in their houses in Hay Street, so someone thought we might as well go the whole hog; legalise gambling, build casinos, put poker machines in hotels, erect lots of motels and hotels ready for the expected influx of visitors, and all have fun. I am glad to say the idea didn't catch on.

The towns have had many ups and downs since that evening in June 1893 when Paddy Hannan left his mates, Dan O'Shea and Tom Flanagan, to ride into Coolgardie and lodge his Reward Claim with Warden Finnerty. The news of the new find, twenty-five miles to the east, sped through the camps like wildfire and caused a stampede of men, all rushing to reach the new field and get their pegs in first. The winter rains had just soaked the countryside, and the going was heavy through the mud and scrub. Some men did not wait till morning, as most did, but dashed off in the night. An early writer stated that he would never forget the sight at daylight of men all streaming in the one direction; some with horses and camels, others pushing wheelbarrows, or many, like himself, heavily burdened with swags tied on their backs.

The whole area, in what was later to become the fabulous Golden Mile, the richest mile in the world, was soon a hive of activity. Some men were fortunate enough to peg good ground and find rich stone and nuggets of all sizes, while others who pegged nearby were not so lucky.

As soon as the earth dried out, which does not take long in this dry country, the men poured earth from one tin pan into another placed on the ground, letting the finer soil blow away each time, till the stones and nuggets were left behind. This is the most primitive form of extracting alluvial gold. The men who owned "shakers," or dry-blowers, shovelled the earth in and began shaking the dirt away. It was not long before they took on the colour of the earth around them.

Paddy Hannan, a quiet middle-aged man, did not like the crowds of men, nor the chaos he had helped to create, so he left for further prospecting in quieter parts, near Menzies.

His bronze statue stands on the corner near the Town Hall in Hannan Street, Kalgoorlie. He appears to the passers-by as if he were offering them a drink from his waterbag. No tourist would dream of leaving Kalgoorlie without photographing the statue or having their photo taken sitting on Paddy Hannan's knee. Over recent years it has been the practice of some New Year revellers to give the statue a coat of paint. In 1969 it needed re-bronzing; the repeated harsh cleaning to remove the paint wore out the bronze. When New Year came around again, a police guard was stationed to make sure the yearly act of vandalism was not repeated.

With hundreds of men working feverishly in the new fields, the alluvial gold became scarcer and harder to find, so men drifted away to other more promising fields.

Some went north, but many went east. Some of these prospectors found gold. Soon the townships of Bulong, Golden Ridge, and Kanowna sprang up. It seemed that Kanowna would take precedence over Hannans when an alluvial field of some importance was discovered. Originally named the "White Feather," Kanowna gained much prominence in 1898 with the output from its deep leads. Old timers say that half of the gold from the deep leads was never reported. The gold was in the form of nuggets of all sizes, so the prospectors from the Eastern States, the T'othersiders, only sold enough gold locally to pay their way and took the rest back East with them.

Following the deep leads, the miners discovered reefs at Kanowna, and several companies were formed to work them. A peculiar feature of mining in the Kanowna district was a "clayey" deposit, known as "pug." Although richly impregnated with gold it formed a difficult extraction problem.

Kanowna, twelve miles from Kalgoorlie, had a fairly long life, but except for shafts, dumps, cleared ground, and markers erected by the Eastern Goldfields Historical Society, as well as two old cemeteries where the early gold-seekers who died from typhoid were buried, nothing remains of this once thriving township where hundreds of people lived and worked.

A depression settled over the field at Hannans, but a few experienced miners saw the reefs of rich quartz, and finally convinced the London sharemarket that their money was worth investing to open up and follow these reefs.

In 1897, ore that was thought to be worthless mullock, and had sometimes been used to build chimneys for the miners' first homes, was found to be sulphide ore which imprisoned rich, fine gold. These telluride minerals are unique and the Golden Mile is the only place in the world where they occur in such variety. A previously unknown variety was named "Kalgooralite." The companies were then faced with the problem of extraction, and when this was finally overcome it marked the beginning of a new golden era.

By 1903, two million fine ounces of gold were extracted. In 1904 the Great Boulder Pty paid dividends equal to $162\frac{1}{2}$ per cent. In those few years, Hannans had grown into the city of Kalgoorlie, with electric trams running to Boulder and Lamington, miles of streets formed, and some handsome homes built. The word *Kalgoorlie* is derived from an Aboriginal name for a silky pear that grows in the district.

Over the years, production gradually declined till by 1919 only 734,000 ounces of gold were produced in the Golden Mile. Mounting costs and a lower gold yield continued after the first World War for a decade, and people drifted away, some shifting their houses to the agricultural area.

One of the twin circular Durkin shafts being collared. This South African technique was used at Kambalda, and provides one shaft for haulage and one for service use

It was during this period that our family (my father, mother, and I, as a baby) left for Victoria.

The world depression of 1929 brought a price rise for gold, and, combined with improved methods of production, made conditions in the mining industry look much brighter, and began another period of prosperity. Mines which had been closed for some time reopened, and huge dumps were re-treated.

Seeing the depression coming, my father decided it was time to come back West to Kalgoorlie, where he realised conditions should be better. We travelled back on the old steam driven Transcontinental train, bringing back with us an addition to the family, my new sister.

Fortunately we were able to rent the same flat in Porter Street that we had occupied before we went away. My father used part of it as a violin teaching studio.

I arrived in Kalgoorlie in time to receive, as a new pupil at the Kalgoorlie Infants' School, the bronze centenary medal struck to commemorate the State reaching its hundredth birthday, and given out to all Western Australian school children.

As I was coming home from school one day, I noticed a large excited crowd in Hannan Street. They were gathered around something, all craning their necks to get a better look. Being young, I was able to wriggle through the crowd and saw, displayed on a table, the wonderful Golden Eagle Nugget, the biggest nugget found in Western Australia. It had been discovered by young Jim Larcombe and his father at Larkinville. One of the men guarding it noticed me staring fascinated at this fabulous, eagle-shaped chunk of gold. He bent down and said that I could pick it up. I tried, but to the amusement of the onlookers, it would not budge. No wonder! It contained 1,135 ounces of gold!

The hotel in Boulder, bought by Larcombe with the proceeds of the nugget, and named the Golden Eagle Hotel, has passed into other hands now, but the Golden Eagle sign still hangs in front.

The Golden Mile managed to hold its own for a few more

Top: A prospector's camp and show near Ora Banda thirty years ago. Bottom: Exploration shaft at Poseidon

years by continually experimenting to find cheaper and better ways of treating the ore and tailings, but once again rising costs, skilled labour shortages, and the pegged price of gold caused the townspeople to be anxious about their future. The optimists said "something will turn up," and it did. Its name was Nickel.

2
THE DAWN
OF A NEW ERA

THE FINDING OF NICKEL at Kambalda was the beginning of an exciting new era. When it was realised that we had other treasures, besides gold, hidden in the ground, the goldfields took on new life, and we felt as if we had experienced a last minute reprieve.

Who discovered Kambalda? I believe that the discoverer of anything is the person who "sees what everyone else sees, but thinks what nobody has previously thought," or who "sees something he believes is significant, and does something substantial about proving it to be so." Several people played a part leading up to the day in January 1966, when a Western Mining drill bored into high-grade nickel sulphides at Kambalda.

The story began in 1897, and the first people involved were early gold prospectors who found rich gold at a place called Red Hill, thirty-four miles south of Boulder. A mine was established, and later Surveyor Roe planned a townsite nearby. He called the new town Kambalda, as it "sounded a pleasant name."

30,000 ounces of gold were mined from 40,000 tons of

ore between the years 1898 and 1906, when the gold finally cut out. In one rich pocket, 550 ounces of gold were extracted from three hundredweight of stone. A rich specimen named the "Golden Butterfly," on account of its shape, was found at what became known as the Butterfly leases, and was much photographed and admired.

Little was recorded of those days at Red Hill, but one old photograph shows men from the town with a land yacht they were trying out on the surface of Lake Lefroy close to the Red Hill mine. It looked a home-made vessel (or vehicle) with its oblong, box-like body on four wheels, topped with sails to catch the wind. This photo could perhaps have inspired others seventy years later, because in January 1970 two modern land yachts were brought up from Northam by enthusiasts to try out. The sport was soon firmly established in what is considered ideal conditions. Some thirty miles long by eight miles wide, Lake Lefroy has about 150 miles of flat, salt surface with occasional islands of scrub and trees. A feature of Kambalda's life nowadays is the brightly coloured sails of modern land yachts, racing on the lake at weekends.

In the eight years of Red Hill's, or Kambalda's life, the outcrops of what turned out to be some of Kambalda's nickel sulphide lodes were discovered and investigated by gold prospectors, who sank a number of surface pits and shafts. The significance of these outcrops was not recognised by these men, and as no other mineral but gold was of any interest to them, the workings were abandoned.

After the mine closed, the place became deserted, except for a few prospectors. All that remained were the grey dumps, rotting timber, and rusting machinery of the mine, a surveyed townsite which had not had time to develop very much, an old well, and two grave markers.

The next part of the story could be said to have begun when two men met and became friends in 1938. One was George Cowcill, and the other John Morgan.

George Cowcill was working on his father's farm at South Quarading when he heard the exciting news that young Jim Larcombe and his father had found the fabulous

Golden Eagle nugget at Larkinville. This inspired him to leave the more mundane work of the farm for a while to join the rush of men going to Larkinville to see if there were more fortunes waiting to be found. Larkinville, named after Mick Larkin, who had discovered the field two years previously, was in country south-east of Red Hill. For six months George Cowcill prospected, mainly in this district and around nearby Red Hill, but with little success. When the little money he had taken with him was gone, he returned to the farm.

Over the next few years, whenever things were slack on the farm, and he was not needed, he returned to the gold-fields, taking on various jobs, working in gold mines, and often carting firewood. While on one of his trips, George Cowcill met John Morgan who was then working in the Celebration Mine. They worked together in two or three of the same mines, and at weekends, the two men often went woodcarting and prospecting together.

During a woodcarting trip, George Cowcill once made an alluvial find six miles west of Woolubar Dam. It was put on the map but mis-spelt "Cossel's Find." He found gold nuggets of various sizes in this field weighing up to twenty-eight pennyweights, and one as much as thirty-two and a half pennyweights. He spent a lot of time prospecting around the Red Hill area, and said, "On one of these trips I had a nickel gossan in my hand and didn't know what it was."

In the 1954 uranium boom, which attracted a horde of prospectors, George Cowcill went out to Red Hill once again, this time with his eldest son, Les, and picked up samples of rock, hoping that they would prove to be radio-active. He took them into the School of Mines for testing, and was told that there was no sign of radio-activity, but the samples did contain traces of nickel. Bill Cleverly, a senior lecturer at the School, wrote this report: "In view of the very meagre Australian sources, and the relatively high price, it might be worth while to explore the extent of the occurence and submit a representative sample for assay."

George Cowcill went back and dug some more samples

deeper down and took another lot to be assayed, but with the same result. There were nickel traces in the rock yet this did not interest him. He wanted to find gold.

John Morgan, who was born in Abbey-Dore, a little village in Herefordshire, England, started his mining career in the coal mines of South Wales at the age of fifteen. He fought in the first World War, and the difficult conditions in post-war England made him decide to migrate to Australia in 1924.

For the first few years in Australia, he worked on farms, then at gold-mining near Meekatharra during the depression years, until he came to work in different small mines in the Kalgoorlie area. It was while working at the Celebration Mine, twenty miles south of Boulder, that he met George Cowcill.

By this time Morgan was married with a family, and he moved into Boulder where he worked as a plant operator in the Gold Mines of Kalgoorlie, a subsidiary of Western Mining Corporation. George Cowcill often stayed at his home while on the goldfields, and they decided that if either of them found anything while prospecting, they would go fifty-fifty, as partners.

In 1964 they read in the newspapers that there was a world shortage of nickel, and this prompted them to collect more samples at Red Hill, as George Cowcill had done ten years before. John Morgan took them to the Gold Mines of Kalgoorlie and asked if he could have them tested for nickel. At first he was laughed at, but when Mr W. Symonds, the assayer, had tested the samples a couple of days later, he said, "There is nickel in those samples, John."

From then on, John Morgan attended to most of the business because by this time George Cowcill owned his own farming property, a few miles from his father's farm, and did not have the time to make frequent visits. However, when John Morgan wrote and told him of the result of the tests, he came up once again and they took more samples to show Roy Woodall, a geologist at Western Mining.

Western Mining Corporation Ltd began as a mineral

exploration company concerning itself with the search for gold, primarily in Western Australia from 1933 to 1954, when the company changed its policy and began to direct some of its exploration resources to the search for economic deposits of other minerals and metals. The first success was the discovery of the bauxite deposits in the Darling Ranges near Perth which resulted in the formation of the Alcola of Australia complex in 1961.

Knowing of Western Mining's interest in base metals, Cowcill and Morgan felt that it was the proper company to take their samples to. Hundreds of samples had been taken by prospectors to Western Mining for investigation and analysis over the years, but they rarely proved to be anything of significance, and the Company made many fruitless inspections of prospectors' "hopes."

After the Kambalda samples were analysed, Mr Woodall got in touch with John Morgan, and asked if he could see where they came from. As George Cowcill was back on his farm at this time, Morgan took Woodall out and showed him the location at Kambalda. John was told that the samples were of interest, and there would be a reward if the partner's information led to anything of significance.

Western Mining was already on the look-out for signs of base metals in the goldfields, and was systematically mapping parts of the country. The goldfields region was recognised as being very similar to large areas of Canada which had nickel, copper, and zinc mines near their famous gold mines. Many geologists wondered why this could not also apply in Western Australia. Therefore Western Mining, at the time still a small company, was searching for just the indications of base metals that were shown in the samples brought by John Morgan.

Mr Woodall told Mr Brodie-Hall, of Western Mining, about these interesting samples, and the existence of an outcrop of what could be a nickel sulphide lode. He suggested that the area should be investigated, and a reserve obtained over a portion of Crown Land. From September 1964 until 1966, systematic exploration was carried out in the Kambalda area, revealing numerous indications of nickel mineralisation. Many of these were distant from where the original samples were taken out. As exploration progressed, a temporary reserve was applied for over the area, and granted.

During the following weeks, Woodall and the team of men working with him became more certain that nickel ore-bodies lay hidden under the ground in this area. This was still at a time when few had any faith in the existence of base metals in the goldfields. Mining men had been conditioned for over seventy years to think of only one mineral, gold, in connection with the Western Australian goldfields, and it was difficult for them to think otherwise.

Mr Woodall needed a diamond drill to find out if his theories were correct. Eventually one was brought from Leonora, where it had been drilling for zinc and copper. In January 1966, the drill struck 8.3 per cent nickel in the first drill hole!

George Cowcill and John Morgan were advised of the Board's decision to give them a $50,000 cash reward, tax free, so each received $25,000 as his share. Their part in the Kambalda story was over.

George Cowcill was able to settle down in comfort, and build a new home, although his prospecting spirit is still very much alive. In his spare time he prospects around the Quarading district and has pegged a few claims for nickel. He hopes something may come of these leases but as he says, "Only time will tell."

John Morgan did something that he previously thought was impossible. He travelled to England to see the six brothers and three sisters whom he had left behind thirty-four years before.

Woodall is still very much connected with the nickel search, as he is now Exploration Manager and Chief Geologist of the Exploration Division of Western Mining Corporation Ltd in Kalgoorlie.

3
KAMBALDA IS BORN

JACK LUNNON WAS the first driller who intersected nickel sulphide at Kambalda. "I reckon I did about ninety per cent of the original drilling in the Kambalda—Saint Ives location," he said. "We had a bad run beforehand and we didn't expect too much to come of it. Quite honestly we didn't expect optimistic results at Kambalda. Our ultimate find was a pleasant surprise."

Since joining Western Mining as a drill helper in 1951, Jack has worked for the company in various locations. The first shoot, or ore-body, to be worked was named the "Lunnon Shoot," and the first shaft was sunk leading down into this. Later, to find a suitable name for the shaft, which is on a high bluff not far from the old Red Hill mine, a competition was held at Kambalda. It was won by Mrs Mary Norman, with her entry "Silver Lake Shaft."

A look-out has been built not far from the head-frame of the first shaft sunk for nickel at Kambalda. It commands a beautiful view of Lake Lefroy, and from there I have seen the lake looking breathtakingly lovely, tinted with varying shades of pink, with wine coloured haze shimmering above it. The colours and moods change with the weather. On a bright sunny day, Lake Lefroy is a dazzling silvery white, while after rain, and with the sky overcast, it is sad, with soft grey shades. Stretches of water show, but most are only mirages.

Salt from part of the lake near Widgiemooltha is being mined and exported to Japan. A few years ago it was thought that the long stretches of flat, salt surface would be used by Donald Campbell to try for the world speed record in his "Bluebird." The lake's surface was tested but proved to be too soft.

Harry Rymer was selected from a number of applicants to be the development foreman in charge of not only the first nickel shaft at Kambalda, but also the first shaft sunk for nickel in the whole of Australia. First, with the aid of 1,000 pounds of explosives and crew of experienced miners, he had to blast away part of the hill-side, then level the excavation. On this surface was built the head-frame of the shaft, fitting shops, winder room, and change rooms, as well as a clearing for a car park and a look-out. Behind the town, and about two miles from the Silver Lake Shaft, the service area was built. This included the ore reduction plant and concentrator, the offices, power house, workshops, store, in fact everything necessary to maintain the mining operations.

Mr J. Oliver, who has been with Western Mining since 1951, was appointed Project Engineer in 1966 and became Resident Manager of Kambalda in 1967. He was trained at the School of Mines in Kalgoorlie, qualifying first as a Mine Surveyor, then as Mining Engineer.

Among some of the earlier problems that Western Mining had to face at Kambalda was the lack of an adequate water supply, telephone services, electric power supply, and suitable roads. But by mid-April 1967 an eight inch water main was completed; thirteen miles long and branched from the Coolgardie — Norseman pipeline at Spargoville. This brought the price of water at Kambalda down from ten dollars to fifty-five cents per 1,000 gallons, and also meant that householders could start growing their gardens.

A power line from the Gold Mines of Kalgoorlie (G.M.K.) power house in Kalgoorlie had been erected by October 1966 and operated as the only source of electricity until Kambalda started generating its own power in August 1967. Power from the G.M.K. can still be used as an alternative service. A telephone service replaced a radio communication system, also in October 1966.

With the greatly increased use made of the graded road between Boulder and Kambalda, it soon became cut up, and travellers were covered in thick, clay coloured dust. Conditions greatly improved when the road was bitumenised. A road connecting Kambalda to the Coolgardie—Norseman highway was later sealed also. It is ironic that while the Boulder—Kambalda road remained unsealed, no traffic deaths occurred, but when a smooth black top road was put down, there were several road deaths. The worst tragedy was when two cars collided head on, killing nine people. The victims included seven members of a family of eight in one vehicle, as well as the two occupants in the other vehicle.

Ore was first produced from the Silver Lake Shaft in April 1967, and treatment of the ore commenced in June 1967. On 15 September 1967, the opening ceremony of Kambalda's nickel project was performed by the Premier of Western Australia, Sir David Brand. A large number of visiting V.I.Ps, some from Canada and Japan, together with Board and staff members of Western Mining and television cameramen and newspaper reporters, were taken to Kambalda from Kalgoorlie by buses and private cars.

A luncheon was held in one of the large workshops near the Concentrator Plant, and a plaque, embedded in granite, was unveiled. It marks the spot where the first Kambalda hole was drilled. Each guest was presented with a white leather folder, with his name inscribed in gold, containing booklets and a commemorative brochure which later won the 1967 Excellence in Lithography Award. They also received an inscribed medallion set in heavy plastic. It was made from the first consignment of nickel concentrates sent to Canada for refinement.

The history of nickel is very old. Early man used implements made from meteoric iron, which usually contains between five and fifteen per cent nickel. It is often found associated with cobalt, and is commonly alloyed with copper, gold, arsenic, and tin. There is a story that this strange, brittle metal, which acted so unpredictably during smelting operations, caused Saxon copper miners so much trouble that they named it "kupfernickel," after the mischievous "Old Nick."

It was officially discovered by Alex Frederick Cronstedt in 1751, and in 1879 a German chemist named Fleitmann found that it could be easily drawn and rolled when mixed with a small amount of magnesium. Nickel is often used to coat soft metals by electrolysis, but its most common use is as an alloy in steel. Its hard, rust-resisting properties increase the strength and toughness of steel, and give resistance to heat, corrosion, and low temperatures.

Between 1966 and 1967, ninety-three tons of nickel concentrates (which look like thick, black soot) were extracted from ore at Kambalda, increasing to 11,723 tons in the period from June 1969 till April 1970.

The ore is transported in trucks from the shafts to the concentrator plant, where nickel concentrates are obtained by crushing the ore to "fines." These are then wet-ground to a powder. The powder, a pulp at this stage, goes through a process of magnetic separation which produces a pyrrhotite concentrate. (This is low in nickel but is stockpiled for possible future retreatment.) Finally, flotation treatment separates the nickel sulphide, pentlandite, from the waste rock samples.

Nickel concentrate is railed to Esperance, and stockpiled for shipping to Japan and Canada for sale overseas. It is also being sent to Western Mining's multi-million dollar nickel refinery at Kwinana, which began production early in 1970.

On 9 June 1970, about half a ton of pure nickel briquettes from Kwinana were shovelled into a furnace at the Mayfield plant of Commonwealth Steel Co. Ltd to produce the first batch of Australian stainless steel. Nickel sent

16

back from the first consignment of concentrates shipped to Canada was used for making Australian coins, and nickel refined at Kwinana was also used in our coins for the first time in 1970.

Drilling in the Kambalda area is continually in operation. To check the results, cores from the diamond drills are taken to Kambalda's Core Farm and stored. They are stacked in trays and painted with their respective hole numbers and footage details, so that their origin is precisely known when they are required for assay purposes by the company's geologists. When needed for assay, the cores are cut into portions with a diamond saw. Some are sent to the company's Belmont laboratory for analysis, while others are kept on permanent record at the Core Farm. These provide an exact record of all drillings made in the area, and after five years of drilling there are over one million feet of diamond drill cores stored at the farm.

During part of the exploration drilling programme, more shoots of ore were discovered. Shafts were sunk in two of these, the Durkin mine, which went into production in October 1969, and the Otter mine, which started in November of the same year.

An unusual method of shaft sinking was undertaken on the Durkin shoot. Two concrete lined circular shafts, each fourteen feet in diameter and 1,000 feet deep, were sunk. One was used for haulage, and the other for service use. Circular shafts are much used in South Africa, so an experienced South African shaft sinker was brought out to Australia and placed in charge of the project.

The Otter Shaft burrows deep down into the side of of a hill. A jumbo-drill used in this shaft was featured in the A.B.C. television programme, "Project Australia," on 9 July 1970. The programme commemmorated the introduction of the micro-wave link, which enables a direct dialling telephone system between East and West. Claire Dunn, a television personality and a star of *They're a Weird Mob*, interviewed miners 1,700 feet down in the Otter Shaft.

Western Mining also announced the discovery of two

ore-bodies in the Saint Ives area, fourteen miles south of Kambalda. Saint Ives, once a gold-mining town, is separated from Kambalda by the eight miles of Lake Lefroy, which like many other salt lakes in Australia, consists of a thin crust of salt covering a considerable depth of soft, deep mud. It was decided to build a road across the lake so that nickel-bearing sulphide ore from the Saint Ives deposits could be brought across to the reducing and concentrator plant at Kambalda. This created quite a problem, as no ordinary bulldozer could attempt to move across the lake without disappearing from sight. Even an average car could not move far without sinking up to its axles.

Consulting engineers were called in and they decided to build an embankment using the actual mud of the lake. This was done by excavating along each side of the road by drag-line machines. These huge machines, each weighing twenty tons, stood on a platform of large timber planks fifteen feet long by three feet wide. As the machines moved forward they picked up the planks they had left behind, and swung them around in front again. These working platforms distributed the weight over an area large enough to reduce their bearing pressure on the mud, and proved effective. When the clay bank consolidated and dried out sufficiently to carry the heavy ore trucks, it was surfaced with sand and gravel. Access roads were built to four sandy islands on the lake, and other causeways were built on the salt surface for exploration drills searching down through the lake bed for the southern extension of the Lunnon Shoot. Pipes were laid under the causeway so that the flow of water and salt to the salt mines near Widgiemooltha would not be affected.

The town of Kambalda, built on the original townsite, is set in pleasant surroundings in the fold of two hills. Attractive local flora includes sandalwood, kurrajong, and several varieties of flowering eucalypts, with the lighter green of the quondong showing up among the darker foliage. During the building of Kambalda, a company rule was made that no trees were to be removed unnecessarily, and that all natural growth was to be preserved.

18

As the homes are being built among the natural bushland, the people of Kambalda will have some protection from the choking dust that plagued the housewives in the early gold-mining towns. In the early years, the lives of the residents in Kalgoorlie and Boulder were made miserable by the smothering red dust storms. The dry-blowers had turned over miles of ground and denuded it of all vegetation. When Hannan Street was hardly more than a track, the ground was powdered to talcum-fine dust by the constantly moving traffic of horse and camel teams. The shopkeepers' wares in their first business premises were covered by thick dust, and, unless it rained, a dust pall always hung in the air.

Mining equipment was powered by steam, and needed tons of wood to keep the boilers' fires going constantly. More timber was needed for lining the shafts and stoping underground. All the salmon gum, mulga, and straight gimlet gum nearby were chopped down, and soon there was little protection from the summer dust storms.

Railway lines were extended further into the bush for the wood-line trains to bring in the wood, so vital to keep the mines working. Two of the "wood-line towns" were Lakewood and Kurrawang. The latter is now used as a Native Mission Centre, but Lakewood was abandoned when the last of the steam boilers changed over to diesel fuel.

Fortunately, many years ago, a law was passed which made it an offence for green trees or shrubs to be removed or destroyed within a certain radius of the Twin Towns. The soil is now bound together by a green belt which has grown again, helped by reafforestation by the State Forest Department. This, combined with the gravelled and bitumenised paths and roads, has helped to overcome one of the biggest burdens the goldfields housewife has had to bear — dust.

However, in the summer of 1969, at the end of the driest year since 1897, with an annual rainfall of less than five inches, we had a taste of what our mothers and grandmothers had to put up with. One of the worst droughts on record twice caused walls of red dust, hundreds of feet high, to descend on the towns. The second storm, in February 1970, was an effect of Cyclone Ingred. As well as causing washaways and widespread damage, it created a strange phenomenon. Huge clouds of dust, swept high in the sky from dustbowl areas, mixed with rain and on reaching the ground covered everything with thick, orange-brown mud. Everything was covered — buildings, vehicles, roads, and trees. Also, as I found out too late, washing on the clothes-lines.

Jean Verschuer was engaged by Western Mining as its landscape consultant at Kambalda and her aim has been to create a town which will not imitate the suburbs, but retain its bush atmosphere. In this, she has succeeded very well. She advocated using indigenous flora almost exclusively and the effect is most attractive. Even the names of the streets at West Kambalda are taken from the local flora. Names such as Spinifex Street, Blue Bush Road, Sturt Pea Crescent, Kurrajong Lane, and Pittisporium Street add to the town's distinctive character.

Kambalda West was built on a flat plain four miles to the west of the original town, as it was found that Kambalda could not extend any further owing to the discovery of new, rich ore-bodies where the town was in the process of being built. The homes at West Kambalda, as in Kambalda proper, are of varying design, and set at different angles. So there is no look of dull uniformity. It is self-sufficient too, having its own shopping centre and school as well as a beautiful new Spanish style motor-hotel which was built at a cost of $2\frac{1}{2}$ million dollars.

The Silver Lake drive-in theatre played its part in giving the town a "different" look by building its refreshment rooms in the shape of a silver dome, which looks from a distance like a space craft or something from science fiction.

The Kambalda workers were variously drawn from wide areas — from overseas and the Eastern States as well as from the mining population of the Eastern Goldfields. There was some industrial trouble over wages and conditions of work for a time, but things settled down after a

DRINK KALGOORLIE SPARKLING
AMBER ALE
DRINK KALGOORLIE SPARKLING
AMBER ALE

nickel bonus of twenty-five dollars per week was granted in December 1969. Since then, the workers have settled down in their Company town, and are kept happy by having most of the city amenities. Even television came to the town just a few months after Kalgoorlie received its first T.V. broadcast in January 1970. The long waiting list for houses shows that there are many more people anxious to live in Australia's first nickel town.

TOP: Double-decker trams which ran between Leonora and Gwalia in the 1900s. BOTTOM: Racing camels with their Afghan riders at Coolgardie for the St Patrick's Day sports in 1896

4
PROSPECTING FOR NICKEL

AFTER THE DISCOVERY of nickel at Kambalda, mining companies sent out their geologists to look for other likely areas. Week-end gold prospectors began to turn their attention to nickel. To assist these men, Western Mining supplied the Prospectors' Association with a set of rock samples for identification. The School of Mines also had an influx of visitors to its museum. They examined the types of rocks connected with nickel, and received help and advice from the staff geologists. Armed with their new knowledge, they were soon ready for the field.

The prospector had to learn to "read" the country, and to see if it was suitable for nickel. Hills were favourite targets, as these may be outcrops of rock which could be worth investigating. He had to discriminate between basic and ultra-basic igneous rocks. In Western Australia the ultra-basics have proved to be the most significant host rocks for nickel. The prospector then looked for signs where the ultra-basics had broken down or weathered into by-products such as magnesite, crysophase, opaline silicate, laterites, and so on.

After any of these rock types were located, samples were taken to an assay office to be tested for nickel. If these showed good results, further rock and soil samples were taken out, often by costeening or trenching. Further satisfactory results led to the prospector pegging a few blocks of 300 acres apiece, which would cost him $160 each to register.

The next step was for him to approach a company and offer his leases for sale — though if his results had been very good the news would have leaked out, and the companies themselves would have approached the prospector. Unless there was an outright sale, a company usually took out an option on these leases. The prices varied, but an average of $1,000 per block was often given and perhaps a share issue in the buyer's company.

Further work was then undertaken by the company. Surveyors and their assistants would be sent out to "grid" the area, which means that the survey teams systematically cleared lines through the bush at varying intervals. The geologists tested the ground in these sections, taking out numbered samples at regular intervals. These samples were then assayed, usually for nickel, copper, cobalt, zinc, and perhaps other base metals. By using this method, the geologist knew exactly where each sample came from, and would make further tests when the samples showed promise.

After finding an interesting area, the company generally decided to bring in percussion or diamond drills to locate an ore-body. If time was running out, the option would have to be extended, which gave the prospector more money. When an ore-body was proven, the option was exercised, and a mutual amount agreed upon. The terms usually depended on the original agreement, but often a further share issue in the buyer's company was given and sometimes royalties on every ton of nickel produced were also paid to the seller.

Bert Skinner was one part-time prospector who changed over from gold to nickel prospecting. With a former prospecting mate, Frank Bray, he set about learning what rock types to look for, how to recognise them, and how to know if the rocks contained a significant amount of nickel. They made frequent visits to the School of Mines Museum, and were helped by the staff geologists: Bill Cleverly and Tony Moriarty. With the assistance of the metallurgist, Ernie Tasker, they evolved a system of chemical analysis for use in the field. This showed the nickel content of a rock in parts per million, and gave an accuracy of plus or minus ten per cent.

One day, while prospecting with Frank Bray and Kevin Erbe, Bert Skinner broke a piece from a likely looking rock and subjected it to chemical analysis. The result was one point five per cent nickel. The men were jubilant, believing that their fortunes were made.

The following day, an urgent assay in Kalgoorlie confirmed the chemical result. However, Bill Cleverly soon dashed their hopes to the ground. The rock is now under glass at the School of Mines Museum and is known as the "Fenbark Meteorite!"

Early in 1968, Bert Skinner and Frank Bray were approached by two business men, Peter and Ted Englebrecht, who asked them if they could assist them in the field analysis for nickel. They joined a syndicate of ten members known as Westralian Nickel and pegged mineral claims at Ora Banda, Siberia, and Canegrass, all north of Kalgoorlie. These properties led to the formation of a public company known as Westralian Nickel Exploration N.L.

After achieving considerable success as a prospector, Bert Skinner was appointed Field Supervisor of the company in 1969, and became one of the new full-time nickel prospectors.

Nickel prospecting was such a lucrative business that other people, besides companies and converted gold prospectors, joined in the hunt. Kangaroo shooters around Leonora became quite wealthy through finding samples of the likely looking rock in country where they had been shooting. They sampled, pegged, and sold their leases for a tidy sum.

Amy Freeman, now Mrs Polletti, a cook working in a hotel in Leonora, was used to hearing nothing but nickel talk all around her. She was interested in rock and gemstone collecting, so one day she decided to join in the nickel search. Taking Aboriginal helpers with her, she prospected in the harsh country around Leonora during her time off from work. Soon she found what she called "some interesting stone," pegged the ground, and sold several blocks to companies.

An Aboriginal dingo trapper who works around the Eastern Goldfields, and naturally knows the country very well, was able to buy a new car and a few other luxuries. He was paid $6,000 for pegging ground for one company, and for a day's work with another he was given $5,000 and 5,000 shares. His day's work consisted of being flown north to an area one early morning, spending all day pegging the site, and flying home again by night.

The Aborigines, both bush and educated, have joined in the search too. They are shown the types of rock the nickel men want, and then go out and look for more and bring in samples. If these samples are what the geologists want, the Aborigines lead them to where the rocks are found, and receive payment. Hundreds of useless samples were brought in, but among these there were some that proved valuable.

It is said that some of the big nickel strikes were originally discovered by the Aborigines. Generally they don't move far from their tribal areas, but, within these, they know practically every feature. So it is not too difficult for them to find the rocks which they think the white man wants. Some Aborigines are pegging claims too. They are often helped by white men or their more educated brothers. Usually, the first thing they buy when they receive enough money is a car.

Another new type of prospector is the pastoralist-prospector. About ninety per cent of station owners in the new nickel fields have pegged at least parts of their properties, as well as some areas that are not their own, for nickel and other base metals. At first they felt compelled to peg portions of their station properties to prevent mining companies from doing so. They felt that the companies were invaders, and could cause damage. However, most pastoralists ended up pegging for the same reason as everyone else: money. With the pastoral industry in the doldrums through falling wool prices, drought, and credit restrictions, they quickly realised that they had found a way to help their stations over a bad period. Some pastoralists have become deeply involved in the mining industry, and at least one is Chairman of a Board of Directors of a mining company.

The first thing a prospector must do is obtain a Miner's Right. It can be bought for fifty cents at any mines department, or at the police stations in small towns. The fee is not expensive when one remembers that Paddy Hannan had to pay one pound for his. A Miner's Right is a prospector's passport to mining, and no claim can be lodged without it.

I have been told that most of these people whom I call "prospectors" should not have that title at all. Instead they should be called "real estate men," because all they are interested in is pegging blocks and selling them. And who can blame them? However, their days are numbered, as there is only so much ground with nickel potential left to be pegged.

My informant also told me that except for trained geologists, there are only about four real nickel prospectors in the area. To be a nickel prospector one would need an advanced knowledge of geology, petrology, mineralogy, and geochemistry, plus a good understanding of the regional structure of the country.

The nickel business is like playing detectives or a big guessing game. The nickel ore-bodies lie hidden, often hundreds of feet below the earth's surface. The clues are on top for the men to find, but there are so many false clues and trails. All the ultra-basics and host rocks *could* lead to nickel but there is no guarantee that they will. Moreover, even if an ore-body is located, it might not be large enough to be a commercial proposition.

As the news of the nickel discoveries in the Kalgoorlie district spread, and some idea of the immensity of the field was beginning to be realised, the world's mining companies sent their leading geologists and exploration crews here. Some of these men found familiar faces when they met each other, as some had worked previously in the same areas, such as Sierra Leone in Africa. Now they were in a new type of country and faced strange conditions. Our extreme weather contrasts of heat and cold had weathered the rocks badly, and made them difficult to identify.

Office space and housing became difficult to obtain as more and more exploration company representatives arrived in the twin towns to set up offices and find living accommodation. Early in 1970, it was estimated that more than 120 exploration companies had set up business in Kalgoorlie and Boulder. They snapped up old buildings, shops, and unlicensed hotels that had been empty for years, renovated them and gave them a coat of paint for use as offices. New office buildings and homes have been built to meet the demands of the increased population.

Companies, as well as some prospectors, are using more sophisticated methods in their search for nickel. The geo-physicists are assisting the geologists by testing the ground with electrical equipment in aircraft. Much of Western Australia is covered by deep soil and laterites, so aerial surveys help to penetrate this blanket and discover what lies underneath.

The government has produced aero-magnetic maps, which outline the subterranean layers of some basic mineral and magnetic rocks. The prospector can examine these at the mines departments, and check to see which land is still available for pegging. With this knowledge, he can go straight to pre-determined areas to do his prospecting instead of wandering through the bush.

Unfortunately, the magnetic method can be disappoin-ting. Some companies, drilling around Coolgardie and Widgiemooltha, have found to their cost that the surveys discovered only graphite slates.

Green-horn prospectors, as the seasoned prospectors call them, joined in the searching and pegging. Enticed by exciting nickel talk and newspaper articles telling of quick riches, they came from everywhere and from all walks of life. Advertisements in mining periodicals and newspapers, including our own local paper, were a big incentive for peggers. Some advertisements were phrased: "Wanted. Nickel Blocks. Near Glamour Areas Preferred. Top Prices Paid." "Option Required of Nickel Leases in Well-Known Areas." One group was so anxious to buy that they offered $1,000 for any mineral claim in known nickel areas without bothering to have the leases investigated.

As the mining fever spread, up to sixty-five Miner's Rights were issued in one day in Perth, and at one stage the Kalgoorlie Mines Department temporarily ran out of permits.

Some of the "green-horns," new to the bush, have been a cause of worry to the people of the Northern Goldfields. They do not realise the danger of being lost in the wild, inhospitable country. Some are secretive about their movements, and wander into the bush without telling anyone which direction they will be going. In 1969 two young men, a geologist and his field assistant, were stranded for more than a week when their utility and caravan became bogged in sand. They were about 140 miles north of Laverton in uninhabited country and had no two-way radio. It was summer, and their food and water soon ran out, but by a stroke of extreme luck they managed to discover a soak where they dug and found water. Luck was still with them when they found an old trap, in which they snared two kangaroos. Their strength was failing as they finally managed to drag their vehicle out of the deep sand, breaking their hydraulic jack in doing so.

Their employers became alarmed because they did not return to Kalgoorlie on Christmas Eve, when they were due. The Laverton police were notified, and when they found that the men had not called into Bandya Station, north of Laverton, as planned, a sergeant and a native tracker began a search. An aircraft was also chartered and on Christmas Day the men were spotted from the air. It was not long before a search party reached them. Except for losing weight, the men were none the worse for their ordeal.

A month later, another party consisting of one man and two youths became lost while they were examining claims for a newly formed mineral company. A high tension lead in the Landrover's engine broke down. "The only spare we didn't take," their leader belatedly realised. This party was also heading north of Laverton in above century heat and the country they were stranded in was barren and devoid of landmarks. They had sufficient food, but water was beginning to run low.

After the men had been rescued, when an air search had sighted them and directed a ground party to the site, District Inspector L. Menhennet said, "Death in the bush could be avoided if people would take precautions before going off into unknown country. They should study the area through which they would be travelling, take sufficient food, water and fuel to last over the trip, and inform the police of their route and date of return. It is important that they should remember to stay with their vehicles." He went on to say, "If they are overdue and a search party is sent out, it will find the vehicles first. There have been too many incidents recently of unwary people leaving their cars and making for highways and homesteads which may be miles away."

In the business of prospecting for nickel there is great secrecy and prospectors go to great lengths to cover their movements. When men go out on field expeditions, they often take elaborate precautions against being followed or having their destination known. One prospector, suspecting that he was being followed on an unused and lonely bush track, hid his vehicle off the road and waited. A little while later a Landrover came along, followed at a discreet distance by a utility. The follower was in turn being followed.

In the few months after Poseidon made headlines, the spying activity increased. Prospectors admitted to deliberately leading spies on false trails. The aerial spies were not shaken off so easily although a man with a good knowledge of the bush often had a lot of fun leading his pursuers all over the country. But it meant wasting a lot of time and petrol.

Blue Gallagher, a veteran gold prospector who worked his gold claim 140 miles north of Leonora, said to a journalist who interviewed him, "A hundred and forty flamin' miles from nowhere but you can't leave your camp in the bush to drop your tweeds without a helicopter spluttering in from over the next hill so that someone can see what you're doing."

However, no matter who the prospector is, or where his leases are located, there is only one way to find out if he or the company who buys from him has a mine. That is by drilling.

5
DRILLING

MOST DRILLING COMPANIES have their headquarters or base camps in Kalgoorlie or Boulder, and each of them owns several drilling outfits: at least two or three, upwards to about eighteen. Some of the larger mining companies have their own drills, but most find it more economical and convenient to contract their programme to drilling companies.

The cost of hiring drills is high. The contractors charge an average of seven dollars a foot for working with diamond drills, but the drills themselves and their spare parts are expensive and costly to run. Diamond-drilling "bits," the components which contain the diamonds and cut through the rock, average about $250 apiece and need frequent replacement. The diamond drill's hollow rods, which penetrate the earth and collect the core sample, are another major item. Different sizes need to be bought to suit the individual requirements of the hirers.

At times the rods become jammed while drilling in difficult country, and hold the progress up indefinitely. If it is impossible to remove the rods, the hole is abandoned.

Both rods and "bit" are lost, together with time and wages laid out for no return.

Except for the small auger drill, which is used by the prospectors for drilling down to approximately twenty feet for surface indications and samples, two types of drills are used; the percussion and the diamond. The percussion drill penetrates fast and easily through broken formations, averaging as much as 500 feet per day to the diamond drill's eighty feet. Also, work with percussion drills is much cheaper per foot. A great advantage is that they give rapid surface indication of where the base rock lies, and they are useful for "pre-collaring;" preparing for the diamond drills. The latter often meet trouble in broken ground, so it is usual for the percussion to be used until the base rock is found, when the diamond drill takes over and bores into it with its hard tip.

The disadvantages of the percussion drills are inaccuracy and difficulty in interpreting results. They only provide a chip sample, and their depth range is limited. The diamond drill, by bringing up a core sample, gives better and more detailed information, and a more accurate estimate of the lode tonnage.

The difference between early gold mining and the present day use of the diamond drill is great. In the early days, drills were not much used until the bigger gold mines began operating. The pioneer miners never knew whether their gold-bearing reef would continue at depth or cut out early. The only way they could find out was to sink a shaft and do a great deal of work before the picture became clear. Often, after a rich "find" showed promise of great wealth and the owner had sold it to a company which then undertook the extensive work to open up the mine, the gold tapered off and most of the company's work and money were spent in vain.

One of the worst disappointments was at Londonderry, south of Coolgardie, where a wonderful "hole of gold" was found in 1894. The owners sold to a company in England formed by Lord Fingall, but after only a few months of work the gold trailed right off. Instead of becoming very rich, the company lost money. If drills had been used on this and other such "finds," the companies could have discovered whether the mine was worth purchasing while it was still under option.

The initial cost of equipment and development is so high on the new mining fields that it is essential to know the tonnage and grade before commencing work. By carrying out a comprehensive drilling programme at the beginning, the company can gauge the life expectancy of the mine. It can then decide whether to build a permanent town, like Kambalda, or to commute its workers from a nearby established town. The type of buildings used on new nickel fields that are too far away for commuting, also depends on the life of the mine. Easily transported houses are more practical for use near mines which have only a few years of estimated life.

Experienced drillers are hard to obtain and to retain, so conditions must be made as attractive as possible to keep the men happy. Their wages are high and usually free board and accommodation are included. And they are not fed on dampers baked in the camp oven, or on tins of bully beef. They expect, and usually get, the best of tucker. Caravans or converted buses are generally used for their quarters, and these contain all the modern conveniences such as gas stoves, refrigerators, air conditioners, and heaters. Their camp sites need to be moved fairly frequently, so it is simpler to use mobile accommodation.

But, even though the workers' comforts are looked after as well as possible in the bush, the labour turnover is still high. To a great many men who work on the drills, it is only a means to an end. Often they work in the bush only until they have saved enough money to leave. Sometimes this only means two or three weeks' work.

Still, if all the spy stories one hears are true, some drillers' days have been livened up with a bit of drama. Once two drillers were working on day shift in an isolated spot when a helicopter landed close by, and out stepped a well-dressed man. Using an authoritative manner, he asked the men how far they had drilled, then picked up the core box and said,

"I've been told to take this back with me immediately as it is urgent." He nodded goodbye, climbed into the helicopter and spiralled off. The drillers thought he was one of the "heads" of the company they were drilling for—but he wasn't, and they still don't know who it was.

Another story is that drillers near Carr-Boyd Rocks gathered white quartz stones and spelled out a message on the ground for the aerial spies to read. They were told to "nick off" but it wasn't expressed as politely as that.

It is well known that during 1969, a syndicate was formed for the express purpose of spying on the nickel-fields and selling information to their clients. These clients, many of whom were brokers from different parts of Australia, as well as from overseas, paid a handsome fee for this service.

A lot of information can be obtained from a helicopter or plane flying over a line of drills. The pilot notices their position and the direction the drills are pointing, which gives some idea of the way the lode is running. He counts the empty diesel drums and estimates the quantity of fuel used and the depth they should have drilled. The colour of the dirt thrown up by the percussion drills and sludge from the diamond drills tells a story too. Some drillers, on hearing a plane or helicopter approaching, often hastily use additives to change the colour of the sludge or earth. One company rid itself of aerial spies for a time by pouring sump oil on the sludge, which the pilot interpreted as a nickel sludge, and flew off excitedly to report. Tales like this moved the stock market.

As many as six planes a day were seen flying over Poseidon when it first held the world's interest. Tales of spying came from all directions about that time. Several stories told of helicopters landing where companies and drillers were working and their occupants dashing out to grab samples. One James Bond type story was of a man who came down in real Hollywood tradition on a rope hanging from a helicopter while the drillers were at lunch. He snatched the core sample, then was hauled back into the helicopter.

Elaborate security measures are often taken to prevent leakage of news from the drillers. One company does not allow the drilling contractors to remove the cores, but sends their geologist to do the job. However, most drilling contractors are very secretive because they know that they face instant dismissal if they are the cause of any leakage of information.

As in the days of the first gold prospectors, water is still a problem. The diamond drill needs up to 5,000 gallons of water a day for a ten-hour shift. If bore water is not readily available, the contractor has to cart water from the nearest source, often for many miles over rough bush roads. The tanks, if only holding 1,000 gallons, need to be filled ten times for the day and night shifts. Some contractors are paid thirty dollars for every tankful carted, depending on the distance. This adds greatly to the mining companies' costs, but it is an unavoidable expense.

6
WATER –
OR THE LACK OF IT

MY MOTHER WAS SPRINKLING water to lay the dust on the unpaved footpath in front of our house one day, when she looked up to see an old man watching her. "Do you know, my dear, men have died for the want of that, and there you are wasting it on the ground," he reproached her. She remembered her mother telling her of how she had paid two dollars for every 100 gallons of water when she and her family first came to live near the mines at Boulder. Water, or the lack of it, was a constant source of worry and frustration to the housewives. The whole family would often have to use the same tub of water for their bath. The cleanest washed first, the dirtiest last, and then perhaps it could be used to throw around to lay the dust.

Water was the gold prospector's biggest problem. It was said that while men were out looking for gold, they spent most of their time searching for water. At first the only water available was from soaks, which are found by digging at the edge of granite outcrops or in creek beds. There is often only a little water seepage, but this was sufficient for the Aborigines who relied on soaks. When the white man

TOP: Men scraping salt into heaps on Lake Lefroy about fifty years ago. BOTTOM: The causeway built across Lake Lefroy from Kambalda to St Ives

KALGOORLIE MINERAL
WESTERN AREAS OFFICE
RESTAURANT
Milk Bar
GIFT SHOP
GARDEN & HARDWARE

came, he would often ruin the soaks by enlarging them to water his camels or horses, and so leave the Aborigines without water. It is not surprising that a few white men were speared in retaliation.

A remarkable feature of this arid country is the number of granite outcrops which appear at intervals. On nearly every patch of rock is a cavity called a gnamma hole which holds water. These holes are nearly always of the same shape, and are situated in the best place for the water to collect. It is not known how they were formed; some say by natives in by-gone ages, while others say they are natural formations caused by weathering processes and chemical change.

Some men gave up looking for gold and built condensers. They decided that they could make more money by selling water distilled from the salt water of Hannan's Lake. A simple type of condenser was made by filling a tank with salt or impure water and lighting a fire under it. A curved length of pipe protruded downwards from the top of the tank. Steam formed in this and dripped out at the other end as pure distilled water. It was a slow and tedious business. Usually a row of such tanks was built, and if the corroding salt was not cleaned out often the tanks frequently burned out.

Later, brackish mineralised water from mine shafts or bores was used in the condensers. Similar water was used in the mining machinery, and caused frequent breakdowns. When water was very scarce, the public had to pay as much as from fifteen cents to twenty-five cents a gallon, and the severe rationing meant that a man often received no more than a gallon a day. The nearest thing to a bath he could have was a wet cloth to wipe his face and hands. Instead of being able to wash his dirty clothes, a good shake had to do.

The tale was told of a man who saw black threatening clouds, expected a summer thunderstorm any minute, stripped, and with soap and a mug-full of his precious water lathered himself all over. He then stood outside waiting for the heavy rain to finish the job. Alas! As often happens on the goldfields, all that arrived was a thick, red dust storm. The optimistic bather ended up like a sticky Red Indian.

The water problem was relieved somewhat in 1896 when the railway reached the goldfields, as water could then be brought in by train. Therefore, one of the most wonderful days in Kalgoorlie's history was the official opening of the reservoir at Mount Charlotte by Sir John Forrest on 24 January 1903. The weather was extremely hot but no-one cared now that there was this abundance of clear water. On the day of the opening the whole town celebrated and everything *except* water was drunk. It was a tragedy that C. Y. O'Connor, the engineer, had not lived to see that day. The Goldfields Water Scheme was his brain child, which turned the idea of pumping water through a 352-mile pipeline from Mundaring Weir to Kalgoorlie into a reality. He died broken-hearted before the scheme was finished, his life having been made intolerable by adverse criticism. Critics said that his idea would not work and was a waste of government money. It did cost $6 million, but it has been worth every cent.

My grandfather's cousin, W. D. Toy, one of the early arrivals at Hannans, once told me, "I remember the day the Water Scheme opened as if it were yesterday. Your grandparents were with me that day. We went up to the reservoir in all that heat to hear Sir John Forrest declare the Scheme open. It was terribly hot that week, every day well over 100 degrees. But what a day that was! Celebrations everywhere. Did you know that a murder was done on that day? A fellow called Ginger Sly was shot dead in the back bar of the Australia Hotel. The man who shot him, a bloke named Kennedy, went to the police station a few days before and told the sergeant that if Ginger Sly didn't stop shoving him around and tormenting him, he'd shoot him. The sergeant knew something of what had been going on, but he told Kennedy not to be silly. He couldn't take the law into his own hands. Everyone was celebrating that day the water came, and Kennedy walked into the bar of the Australia and ordered a drink. Ginger Sly was up at the other end of the bar, but as soon as he saw Kennedy he walked over and gave him a shove and taunted him about stealing his girlfriend from him. Kennedy pulled out a revolver and shot him dead. He couldn't have been charged with actual murder as he didn't spend too many years in gaol." Toy went on to tell me, "I was on my way to Sydney in 1908 to see Jack Johnson beat Tommy Burns in the world heavyweight boxing championship. I stayed in Adelaide for a while on the way over, and one day I walked into a hotel in King William Street to have a drink, and who do you think the barman was who served me? Kennedy!"

Kalgoorlie and Boulder have no appearance of desert towns now. There are parks, grassed playing fields, shady streets lined with graceful gums, which have replaced most of the original pepper trees, and nearly every home has its lawn and a garden of flowers, shrubs, vines, and fruit trees. We also have an Olympic-size swimming pool and several privately owned pools. The community's first swimming pool, besides a privately owned one in Boulder, was the Municipal Baths which is now used as an ornamental fish pond in Victoria Park. None of this would have been possible without C. Y. O'Connor's Goldfields Water Scheme.

On 3 April 1970, a ceremony was held in Kingsbury Park to mark the phasing out of the last of the original steam pumps that helped pump water on its long journey from Mundaring Weir to Kalgoorlie. Electric pumps are now installed. A plaque and a bust of C. Y. O'Connor were unveiled by the Minister for Works and Water Supply, Mr Hutchinson. The bust, erected by the Kalgoorlie Lions Club, is a replica of the one at the beginning of the pipe-line at Mundaring Weir, and can be seen in Kingsbury Park, near the swimming pool.

Mr Hutchinson said in his speech that multi-million dollar plans were being prepared to meet the increasing demand for water by mineral developments in the goldfields. Depending mainly on the size of the nickel developments, the plan could involve the expenditure of more than $45 million over the next ten years. The present average consumption of water in the Kalgoorlie region is six million gallons a day, but it could well rise to ten million gallons. He also said that satisfactory underground water supplies in the goldfields would save some of the heavy capital costs facing the government and industry. With this in mind, a hydro-geological survey would be carried out between Kalgoorlie and Wiluna, where artesian basins are known to contain a great deal of water.

A firm recently demonstrated a new de-salination plant in Kalgoorlie which could be of great help to mining companies, as well as pastoralists, in areas where salt or heavily mineralised water is available. It is also envisaged that in the future, if there is enough industry to warrant atomic power, sea water from Esperance could be de-salinated and used.

When the second commercial nickel ore-body was discovered by Metals Exploration at Nepean, their water problem was relatively easy to overcome by diverting water from the Goldfields Water Scheme on its way to Norseman.

7
NEPEAN

In 1967, METALS EXPLORATION N.L. of Melbourne became interested in the search for nickel in Western Australia and applied to the mines department there for a prospecting area of 100 square miles over the Coolgardie district.

Unfortunately their application was just too late. It arrived on the same afternoon as one submitted by Anaconda. The area applied for was the same, with the exception of an extra corner of ten square miles requested by Metals Exploration to the south of Coolgardie. The Mines Department informed Metals Exploration that they had just missed out on the larger area but that the ten square miles was still available.

Thinking it better than nothing, the company took up the ground and exploration work began soon after. The geologists selected a zone and conducted induced polarisation tests which indicated magnetic ore-bodies. But drilling proved that these were only pyrites and black shales.

This zone was temporarily abandoned and exploration work was directed further away. However, the drill was later moved back to the original area because the geologist, Jim

Couper, still felt sure that they would find nickel in spite of the earlier disappointing findings. The driller, Dick Noble, was anxious to strike nickel too, as he was told that if nothing was found within a certain time, he would be returned to Queensland, and he was happy in the West.

He was lucky. Before long, his drill core showed there was nickel at 900 feet and soon the nickel mine of Nepean was born.

The original Nepean was an old gold mine about a mile away from the new nickel mine, but although a great deal of work had been done, very little gold came out of it. Some of the other small mines or "shows" nearby in the early days were the "Lady Bell," "Iron Ridge," "Golden Scent," "Queensland Extension," and the "Monte Carlo." None of these names have been remembered, so they could not have been anything out of the ordinary.

A different story was linked to the wonderful "Hole of Gold," the Londonderry, eight miles away. At the time of its discovery in 1894, phrases like "A mountain of gold," "One of the Wonders of the World," and "The greatest discovery that Australia has ever known," were published in the newspapers of the time. Unfortunately for its purchasers, it was little more than a hole of gold, because it did not continue at depth.

There is still some activity in a pegmatite dyke near Londonderry. The mine has been open-cut for over forty years and its owners estimate that it could last for another 200 years or more at its present pace. Western Australian Felspar, an offshoot of Australian Consolidated Industries, employs five men to work in its glaring white quarry. Their job is to extract several white stone minerals, principally felspar, quartz, beryl, and petalite, the mine's mainstay. Petalite has a smooth straight grain which minimises heat absorbtion and is used in cleaning fluids, for producing fertilisers, and for manufacturing crockery. This mineral is exported to many countries. The Londonderry mine is claimed to be the only workable deposit in Australia.

Although the stone is mined by modern mechanical means, one job still has to be done by hand. This is the

separating of the felspar, petalite, beryl, and other white
stones from the heaps of rocks dumped from the quarry.
It is difficult for an inexperienced person to distinguish
between felspar and petalite as they look much the same.
Felspar is used in glazing and beryl is a hardening agent
as well as being used in the manufacture of lighting
materials. When picking over the stones some pretty
colours, mainly pink and mauve, are found and they look
very attractive in rock gardens.

Nepean is fourteen miles south of Coolgardie and
motorists driving down to visit the town are halted by large
signboards with bold lettering stating: TRESPASSERS WILL
BE PROSECUTED and PRIVATE PROPERTY, KEEP OUT. Permis-
sion is needed from the company before entering their
property. When I asked why this was, Mr McGann, the
Manager of Metals Exploration in Kalgoorlie, explained,
"After the news got out that we had made a nickel strike in
late March 1968, we were discovering men hiding practically
behind every bush and running their own surveys and
magnetometer tests. This would help them to identify other
areas with similar characteristics. We were absolutely
plagued with curious sightseers and some went as far as to
examine the drill cores. That's why we put the notices up."

When the manager of the new nickel mine, Frank
Lubbock, arrived in May 1969, Nepean was only a tent
settlement, but before long a transportable town was
established to house about fifty single men and six married
men and their families. In 1970, it was reported that the
expected lifetime of the mine would be four years and in that
time $12 million worth of nickel should be extracted from
the ore.

Nepean just beat Scotia to become Australia's second
nickel producing mine on 15 January 1970 when it carted
its first load of nickel ore to Kambalda for treatment.
Being a smaller mine, it was more economical and convenient for Nepean to sell its ore to Western Mining for
treatment at their concentrator plant at Kambalda, thirty-
four miles away. Each truck load is tested and assessed for
its nickel content. A new road was cut through from

Nepean to Kambalda for the use of ore-trucks employed to transport the nickel ore.

Nepean is set in pretty bush surroundings and often about half a dozen kangaroos used to visit an old Norseman personality, Rube Taylor, at the Power House. They became quite tame with Rube, who regularly fed them, although the bush was still their home. Unfortunately, thoughtless shooters have frightened off the 'roos and since then they only come out at night.

At the end of Nepean's life, the whole town will be packed up. The offices, living quarters, canteen, and so on will be loaded onto transporters and moved to wherever they are next needed. Some of the buildings could perhaps finish up at Mount Keith, eighty miles from Wiluna, where Metals Exploration is managing the mine, as well as having a one-third share in the project.

In September 1970, a quarterly report put out by Mr R. Hare, Chairman of Metals Exploration, said that the joint partners Australian Consolidated Minerals, Metals Exploration, and Freeport Sulphur have probable reserves of 120 million tons of open-cut ore assaying 0.55% to 0.60% nickel. Six drills, working for months on the leases near Mount Keith, outlined the ore-bodies and their nickel content.

Metals Exploration, acting on behalf of its partners, bought leases at Mount Keith from various prospectors. Some leases were obtained from Jim Jones, of Mount Keith Station, who had used percussion drilling, the results of which interested the companies. However, two leases nearby, numbers 130J and 131J, have a personal interest for our family because they were pegged by my husband's brother, Harry King, and sold to the joint companies. These leases cover the main ore-bodies of what could become one of the world's giant low-grade, open-cut nickel mines. At prices listed in the early 1970s the ground contains nickel estimated to be worth more than two billion dollars. Unfortunately the prospectors did not ask for royalties when they sold their leases.

8
COOLGARDIE

COOLGARDIE, BEING ONLY fourteen miles from Nepean, is regarded as "Town" by the miners, and the men travelling in and out for shopping or entertainment have helped to liven the "Old Camp." But other factors have helped to revive Coolgardie and make it realise that it could live again. One of these was the pegging of the townsite for nickel.

Late in 1969, a syndicate comprising two local men, Ross Lightfoot and Mario Epis, and P. Malouf, representing a company in Sydney, pegged the only area around Coolgardie left to peg—the townsite itself. A legal wrangle followed with the Coolgardie Shire Council and Anaconda, which held a big temporary reserve in the Coolgardie area. Eventually, the syndicate was granted the claims on stringent conditions. Lightfoot had to sign a $100,000 bond for each of the six mineral claims within the townsite, to guarantee against property damage.

This was not the first time that part of the townsite of Coolgardie had been pegged. On 23 April 1896, Messrs Pearce, Bonner, and F. C. B. Vosper, the latter a journalist and brilliant orator, pegged a gold-mining lease within the town boundary. The area included the lock-up, railway station, part of the hospital, and the Warden's paddocks. Other parties, including Graves, Gifford, and Dinny, followed suit by applying for more leases within the townsite.

Like so many goldfields towns, the founding of Coolgardie, twenty-five miles west of Kalgoorlie, was due to a chain of events in which several men played a part. Between 1864 and 1866, a ticket-of-leave man, attached to one of C. C. Hunt's expeditions which had been sent out to open up a string of water holes in the country east of Southern Cross, discovered gold while on a mission away from camp. He was unable to take up a lease, or even claim the gold, so he hid his find, intending to return when his sentence expired. He died before he was released, but had confided his story of the gold to a fellow prisoner.

Several years later, in 1887, Arthur Bayley of Southern Cross gave hospitality to a prospector, Gilles A. McPherson, and his Aboriginal assistant, who arrived back in the town in a distressed condition after a prospecting trip into the interior. Yarning around the camp fire that night, McPherson told Bayley that he had followed Hunt's tracks because he had heard stories that someone in Hunt's expedition had discovered gold. When about 130 miles east of Southern Cross, he found gold himself, but was forced to return as the country was waterless.

About two years later, at Christmas 1889, another prospector named George Withers returned from a prospecting trip and the story he told William Ford was similar to McPherson's account. He had found gold in the desert east of Southern Cross, but had been speared in the shoulder by hostile natives and forced to return.

Bayley and Ford, both experienced prospectors, had been mates for a long time, but it was not until another couple of years had gone by that they exchanged confidences of the stories told to them by McPherson and Withers. They decided to try to locate the source of the gold.

The two men left Southern Cross in April 1892 and picked up Withers' tracks near Mount Burgess, then followed them to what is now known as Coolgardie. There they found what they had been looking for—a rich reef of gold.

Other men could have been the first to discover Coolgardie, but the first of these did not live to tell the tale. In the rush of those who came after Bayley and Ford's find was one called Jim Moon, who made a gruesome discovery. Some distance from his camp he found pegs in the ground, showing that a claim had been pegged some time before, and one of these bore a tin notice which said in faded writing that someone called Ansden, or Amsden, on behalf of Scott, had pegged the ground in 1888. In a gully close by, the skeletons of two white men were found with their axes lying beside them. These were probably the remains of the first men to peg a claim at Coolgardie.

For years the tin claim notice was in the possession of Jim Moon's friend, Dick Fagan, but it was eventually stolen from the Commercial Hotel in Kalgoorlie.

Other claimants to the title of being the first to discover gold at Coolgardie were F. W. Rudd and Alex Glass, who pegged a claim on behalf of the Anstay Prospecting Company in 1889, although for some reason they did not register their claim at the Warden's Office.

Possibly the strongest contestants to the title were three young men named Harry Baker, Tom Talbot, and Dick Fosser. Having little to do in Southern Cross, as the miners were on strike at the time, the men followed Bayley and Ford's tracks and began prospecting in the same area. There were conflicting stories afterwards as to what happened but the young men claimed that they discovered the cap of the reef which was to become the famous Bayley's Reward Mine. They were not sure if what they saw glittering through the rock was gold or "fool's gold," so the new chums decided that they would show the stone to Bayley and Ford.

Next morning the young men found twenty-four acres pegged around the reef in Bayley and Ford's names and when they protested were advised to peg south of the claim, which they did. Later many people asked them why they were not the first to peg Coolgardie, but Talbot replied simply, "The reason is that we were only lads and knew practically nothing about mining or pegging ground."

The other party flatly denied this story, claiming that they found the reef first. Whatever the truth, the fact remains that Bayley and Ford lodged their Reward Claim, the news of which emptied Southern Cross overnight and started a stampede of men heading east.

At first they came from Southern Cross and York, and then, as the news spread, from other Western Australian towns, from the Eastern States, and even some from overseas. Two hundred men from all walks of life were given a rousing send-off from York as they set off for the new find. A couple of enterprising gentlemen named Cohn and Murphy organised the wagons to carry the men's swags at ten dollars apiece. Many men loaded on their swags and walked beside the wagons. They became known as "swampers" and they soon learned to walk ahead of the wagons to avoid the choking dust. Their journey to the goldfields was difficult and painful. One writer described their hardships as they struggled through the desert: "A common sight at the end of the day was men washing their feet with some of the precious water, and rubbing them with fat or vaseline to ease their soreness. Some had so many blisters they discarded their boots and had their feet encased in sacking." Others pushed makeshift wheelbarrows or carried their own swags.

Coolgardie was named after a slightly altered version of the Aboriginal word, Koolgoorbiddie, given to the Warden's water-hole. Soon it grew into a town of humpies, bough-sheds, and tents, but these were exchanged for hessian and iron buildings. Many disastrous fires raged through the town, with little or no water supply to put them out, so the main street buildings were quickly replaced with stone and brick.

In the few brief years of glory before Hannans, or Kalgoorlie, robbed it of its importance, Coolgardie was a town larger than life. It abounded with adventurers and exciting people. The constant discoveries of gold nuggets evoked almost continuous celebrations, often in the form of champagne parties.

When women began arriving, the unclothed Aborigines who camped on the fringes of the town were considered an embarrassment, so the police collected old clothes from the townspeople and gave them to the Aborigines. They were told to wear at least one garment when in town. The story goes that at a Saint Patrick's Sport Meeting an Aboriginal girl arrived in Coolgardie dressed with a man's paper collar round her thigh, and on another occasion, an old man was seen wearing nothing but a smile and a woman's stocking.

Once again the lack of water was the miners' main problem. C. F. Young, one of those who arrived in Coolgardie in its first years, wrote, "For the first time in my life I paid for a drink of water. We filled our bags at a condenser in the main street — a shilling a gallon."

The first man to sink a well which produced fresh water in Coolgardie was Martin Walsh. The water was a little brackish but it was suitable for stock and domestic use. He made good money for a while till others put on three shifts of workers and rushed their wells down close by. Walsh charged ten cents for a horse and fifteen cents for a camel to drink at his trough and the animals were lined up from dawn till dark.

In the Coolgardie *Courier* on 10 November 1894, this advertisement appeared: "Mr David Long has erected some showers and plunge bath in Sylvester Street at the rear of Stoddart's stables. This is a luxury few will resist, and they can have fresh or salt, hot or cold, shower or plunge, for a moderate charge. Three and a half gallons used in each shower and all the fresh waters saved for various purposes to which it may be put. As the little hymn says 'We have been there and still would go'."

The "all the fresh waters saved for various purposes" is a bit vague but perhaps it is explained in another article I read. After finishing a fresh water shower, a man complained to the proprietor that the water had smelt of perfume and he had discovered a long, blonde hair on himself. The owner of the establishment replied that the water had been first used by the town's new barmaid for a bath, and instead of complaining, he should have been charged extra for the privilege.

The unsatisfactory water supply and lack of sanitation in the first three or four years of Coolgardie's growth caused a typhoid epidemic. Until doctors and nurses came to help alleviate their suffering, the stricken men stayed in their camps, or were looked after by an elderly man in the hastily erected hospital. Sister Margaret O'Brien and Nurse Millar arrived to open a private hospital in Coolgardie, but they deferred the idea for a while when Warden Finnerty called on the women and asked if they would superintend the government patients for the time being. This they agreed to do and Sister O'Brien said, "The dirtiest looking objects imaginable covered in sand and dust indescribable, met our sight the morning after our arrival. Tents, tents and still more tents, men everywhere, old singlets and trousers were the order of their dress. The hopeless look of everything would have discouraged the stoutest heart. However, we got hold of a carpenter to put up frames for our tents and after a week of difficulty, we began our work."

Later they resigned from the Government Hospital, owing, it was said, "to some remarks which were passed," and intended devoting the whole time to their private hospital, leaving the patients at the mercy of untrained nurses. This was the only hospital on a field of 20,000 men.

Soon other women came to nurse the sick and dying and as well as the Government Hospital, several private hospitals opened. Their work was very much appreciated by the men and an item in the *Courier* in 1884 said that "a ring with five diamonds was presented to Nurse Warren at Miss Jefferson's private hospital by the Adelaide camp, for nursing their mate, Jack Sullivan."

To raise money for the Government Hospital, a Ball was held in 1895. It must have been a robust affair, because some of the gentlemen present were soundly censured by a reporter who wrote in the *Courier* that: "The Hospital Ball took place at the Theatre Royal on Thursday night. The attendance good, behaviour bad by some of the so-called

50

upper ten of Coolgardie. A great many excuses can be made in a place like this for men who gaze too long and too lovingly on the yellow fluid when there are only male members of humanity present, but no excuse will cover the conduct of men who have been educated sufficiently well to know better, who use bad language, shout, yell, and cause pandemonium in the presence of ladies. At one stage, things got so badly mixed that a police officer had to threaten to find a different sort of ball-room for more than one of the revellers.

"It is to be hoped that at all future gatherings of this kind, these persons will remember that Coolgardie is *not* a horse-paddock and that the ladies of Coolgardie are as much entitled to respect as their sisters and sweethearts in Adelaide, Melbourne and Sydney, and these remarks will also apply to the imported element, who, although they may have brought their money here, forgot to bring their best manners with them."

In spite of all the devotion and care given by the doctors and nurses, many hundreds died of typhoid on the new goldfields. The worst years were in 1894 and 1895, and the disease was no respecter of persons. Among the victims were the son of Mr Shaw, Coolgardie's first Mayor, and the son of Mr Tucker, the Town Hall Clerk. Even Doctor Ellis, who was treating the patients, was stricken, but survived. The Reverend Collick, the first Church of England minister to arrive in Coolgardie, helped to bury five or six men a day when the epidemic was at its height. Many tragedies were revealed to him by men who had hoped to make a fresh start on the goldfields. Titled men, clergymen, doctors, ex-army and navy officers, and the sons of bishops, found their last resting place in the Coolgardie cemetery. They were buried in coffins made from packing cases and marked "This side up with care," "Stow away from boilers," and "Keep in a cool place." Most were buried in unmarked and unnamed graves. Such a host of men flocked to the fields that many were dead before their names or the names of their next of kin were known to the survivors. Many of them were from the eastern states and their relatives waited hopefully, but in vain, for their return with a promised fortune.

The lure of gold claimed a high price in youth. Most of the men who died were under forty years old, and many of the newcomers were unprepared for what the goldfields had in store for them. They suffered other illnesses, besides typhoid, which were caused by the filthy and chaotic conditions of the goldfields. Dysentery was caused by tainted water or food, and many endured misery and agony of "Sandy Blight," an eye disease spread by flies which had come in contact with rotting animal flesh. Even their diet was a hazard, because the lack of fresh meat and vegetables occasioned scurvy or boils.

The miners' basic diet was billy tea, johnny cakes or "damper" made from flour, baking powder, salt, and water, and tinned meat. Sometimes there were complaints that the Western Australian flour was too "new" and refused to rise. This was an annoyance, but real trouble came from the tinned food. It soon became tainted if it was not consumed at once and caused many cases of food poisoning. The myriads of swarming flies took care of that.

Just the same, tinned meat, or "tin dog" as it was mainly called, must have agreed with some. One night in one of the first hotels in Coolgardie, the Marvel Bar, a crowd of men celebrated the birth of the first set of triplets born in the town. They took up a collection and presented it to the proud father. In replying to the good wishes and the presentation, his reply embraced them all: "Beat that on tinned dog!"

When the Reverend Trestrail's wife had her first child, a rhymed birth notice was put in a newspaper by the proud parents who compared the newly arrived baby to a star fallen from heaven into the lap of its mother. "Smiler" Hales, who conducted the *Mining Review*, came back with:

> *The angels left no gate ajar,*
> *Your wife picked up no falling star,*
> *Whatever you might do or say,*
> *It came on deck in the same old way.*

TOP: Miner's camps clustered around the mines at Boulder in the late 1890s.
BOTTOM: Home units of flats in the Kambalda East residential area, with rock garden and pond in the foreground, illustrate today's mining community.

Father Tracey was the first Roman Catholic priest in Coolgardie and was a fine mixer and a good boxer. He was succeeded by Father Duff, who came from Galway, and was said to have a brogue "so mellifluent you could eat it with a spoon." Arthur Reid tells this story of him: "On Christmas Day, 1894, a crowd of men were playing two-up under the big tree opposite McDonald's saleyards in Bayley Street. The stakes were high and when there was about two hundred dollars in the "kitty," the Father, who was hovering on the outskirts of the crowd waiting for the right moment, suddenly sprang into the clearing, picked up the bag containing the "kitty" and put it in his pocket. He shook his finger reprovingly at the players and said, "Boys, I'm ashamed of you. Playing two-up on Christmas Day!" He used the money to help build the first Roman Catholic Church in the town.

The camels, with their Afghan owners, played a big part in the early years of the goldfields for all forms of transport, although horses of other teams hated to encounter them. Some men must have also disliked the camel's owners, because an Anti-Afghan League was formed in Coolgardie.

Two well respected brothers, Faiz and Tagh Mahomet, controlled the business activities of the camel transport business in Coolgardie, but Tagh was shot dead while at prayers in the local mosque. His murderer, Goulman Mahomet, was caught and later executed at Fremantle.

During the celebrations for the arrival of the railway in Coolgardie in March 1896, it was a toss-up between the natives and the Afghans as to which group was the more eye-catching in the parade. There had been a lot of dissension among the members of the organising committee over the arrangement of the big procession and some had even objected to the Afghans being included at all.

The question was eventually settled and the Afghans, turbaned and wearing their gaily coloured flowing robes, rode their splendidly decorated camels. The natives in the parade were naked except for light muslin loin-cloths.

When the alluvial gold became scarce around Coolgardie, men struck out in all directions and faced the danger of

TOP: The first Warden's Court at Kalgoorlie in 1894. BOTTOM: The Administration Office of Western Mining's Kambalda Nickel Operations, landscaped with native shrubs and trees

getting lost in the merciless desert. In 1893 several men died of thirst after the ill-fated Siberia rush, and the Mines Department began the practice of keeping a fire burning on top of Mount Burgess to enable men stranded or lost to return to civilisation. Prospector F. W. Bow said that he and his mate, Collins, were assisted by this fire several times when night overtook them while out prospecting. Now another type of signal on top of Mount Burgess is in the form of one of the micro-wave towers which stretch from the west to the eastern coasts of Australia enabling us to have a direct dialling telephone system as well as giving us television relayed from Perth.

Although Coolgardie declined in importance when the rich mines opened up on the Golden Mile, it remained an important town for many years. My aunt told me that when she was a girl, it was *the* place to go for an outing. The park was beautiful, with little summer-houses and a zoo of strutting peacocks, some kangaroos, and emus. Victoria Park in Kalgoorlie still had a little "zoo" when I visited it as a child. Where there are now flower beds backing on to the adjacent basket-ball courts, there used to be wire fences to keep kangaroos, emus, and, my favourites, little guinea-pigs in. The big cannon, which aimed at all who walked in the main gates also disappeared from the scene.

Coolgardie did not have mines which had a long life like the mines on the Golden Mile. However, some mines in the district supported small townships and ran for quite a few years. Among these was Burbanks, nine miles south of Coolgardie. Its mines, The Lady Robinson, The Main Lode, and The Birthday Gift kept the town going for about twenty years, only to close at the end of the first World War. There were also the Tindalls mine close to Coolgardie and Bonnievale, a few miles out. In March 1907, Bonnievale was the scene of an accident in which a miner named Varischetti was entombed in the mine for nine days. He was finally rescued by two brave men who used diving equipment brought to them from Perth by a special train.

Three small mines some seven miles south-east of

Coolgardie were worked successfully in latter years. They were The Surprise and Lady May leases found by Mr Baker in 1939, and the Barbara lease prospected by Messrs Cash and E. Scahill in 1944. The Barbara was discovered by Sam Cash using his loaming method.

When the last mine, Bayleys, closed down as late as in 1964, it was decided to turn Coolgardie into a Ghost Mining Town. Ben Prior started things off by establishing an open-air museum of relics from the early days in a vacant block near his garage. It has been a great attraction to tourists and travellers passing through. He even studded the cement inside walls of his garage's "His" and "Hers" with gemstones found in the district but this idea was spoiled by some people prising pieces out for souvenirs, even though there are plenty of gemstones available for sale in the town.

The Tourist Development Authority, in association with the Shire of Coolgardie, turned Coolgardie into a Ghost Mining Town in 1967 and began by placing markers at all places of interest. Most of the markers include photographs and give a short history of the site indicated. The old Court House building, made of attractive, pink-tinted quarried stone, now houses a museum and contains many interesting displays and items. It also gives a comprehensive picture of life in the early years.

To further foster the "Ghost Mining Town" image, the Shire of Coolgardie asked the Railway Department to endow the seventy year old railway station to the council. The council would then be responsible for the upkeep of the station and its facilities. The railway station is now only of historical importance. The recently built standard gauge railway line by-passes Coolgardie. The council plans to have a steam locomotive wait at the station for added interest. An old steam pumping unit from the Goldfields Water Scheme, which was replaced by diesel pumps, has also been given to the Coolgardie Shire to add to its collection of relics of the past.

Ironically almost as soon as the idea of the town becoming a tourist attraction was born, more nickel was discovered in the Eastern Goldfields and Coolgardie gradually began to shed its Ghost Town image. Three exploration companies set up their offices in the town and bought buildings and houses for their personnel. It was not long before every empty house in the town was occupied and new ones had to be built to cope with the growing demand for housing.

As more nickel was found within its boundaries, the Coolgardie Shire, which covers an area of 11,073 square miles, increased in importance. Kambalda, Nepean, and the new finds around Widgiemooltha are all in the Shire territory, and the need for two new Administrative buildings, one at Coolgardie, and the other at Kambalda, soon became a necessity.

The ghost town image may vanish altogether if the leases pegged by Ross Lightfoot and his partners develop into a mine. The leases, which contain old gold workings about half a mile north-east of the town, have been sold, although the syndicate still retains a ten per cent interest. Some drilling on the leases has already given encouraging results. A spokesman for the major shareholder, Chipatoojah Java Rubber Company, promised the Coolgardie Shire Council that if a mine was developed, a percentage of the profits made would go towards the provision of amenities for the town.

Coolgardie will most likely become an important nickel town, but it is doubtful if it will ever recapture the glamour of its past or create the same excitement as the day Bayley and Ford rode from there into Southern Cross with their saddle bags full of gold.

9
SOUTHERN CROSS

SOUTHERN CROSS has not escaped the nickel boom. Geologists are hopeful of a nickel strike somewhere in the district or in the Yilgarn area, where a lot of exploration companies are working. There is a large stretch of granite country lying between Coolgardie and Southern Cross which is of little interest to the nickel men. However, Southern Cross is included in some potential areas which, in common with most promising nickel belts, run in a north-south direction.

Pegging has been very active in the Yilgarn and one group of leases at Trough Wells, thirty-five miles north of Southern Cross, created some excitement on the Stock Exchange in late September 1970. Pegged by International Mining Company, the discoveries gave the share market a much needed boost for a short time when the company announced, exactly a year after Poseidon's historic report, that they had nickel values of up to 3.75 per cent from percussion drilling on the Trough Well project. It was thought a good omen that the good news was released on the anniversary of Poseidon's report.

The company's chairman was a Sydney woman, Mrs Millie Phillips, who was referred to by newspaper men as "Mrs Nickel" or "Thoroughly Modern Millie." One newspaper headed its report of the Trough Well's results with "Joyous welcome for new Queen of Nickelmania — The King (Poseidon) is dead . . . Long live the Queen of Nickelmania."

Mrs Phillips' company was considered a strong contender for taking the title from Poseidon, but unfortunately not long after the encouraging report, there was a drastic mark-down in I.M.C. shares when another report was issued by the company. The drop in share values was attributed to the fact that London was still smarting under the Tasminex fiasco and had over-emphasised the poorer parts of the Trough Wells report, which generally was considered to be encouraging.

The existence of nickel in the Trough Wells district was first reported in mining journals in the early gold mining days, although no interest was shown then. Gold miners mainly considered nickel mixed up with gold a nuisance because it interfered with smelting operations.

Gold was first reported in the Yilgarn (the native word for white quartz) in 1887, when it was discovered at Eenuin by Greaves, Payne, and Anstey, and in Golden Valley by Colreavy and Higgins. Thomas Risely had heard from his friend, Greaves, of the gold find at Eenuin, so he and his mate, Mick Toomey, decided to go there to prospect. They were not successful. One evening they asked Charles Crossland if he could put them on to anything worth prospecting. Crossland was then surveying the claims applied for at Eenuin and Golden Valley.

He pointed to the constellation and said, "If you make in a direction slightly to the east of the Southern Cross, you will strike a line of hills, where a native told me 'Plenty feller gold.'" They followed his directions and on New Year's Day in 1888 found gold, pegged their claims, and called their find "The Southern Cross."

C. C. Hunt had originally named the line of hills nearby the Southern Cross. He was given the task of sinking and enlarging wells, enlarging water catchments, and mapping the country further east, recently taken up by the Hampton Plains Syndicate. While his party were resting at Koorkoordine Well, they noted a range of hills to the south. Hunt also noticed that at a certain hour in a certain month (this was possibly midnight in March 1865), the major axis of the constellation was vertical over the range. When he consulted his nautical almanac the next morning, he said to his assistant, "Yes, that was the Southern Cross."

Sir John Forrest knew of Hunt's original name for the hills, and, coupled with the name the prospectors gave on their leases, he called the new town Southern Cross. The growth of the new mining town was phenomenal and within four years was declared a municipality. After gold was discovered at Coolgardie, it also became important as a staging camp and base from which the prospecting parties left with their provisions. For some time, it was the district's "railhead," until the line was extended to Coolgardie, then to Kalgoorlie and Menzies.

The town was very well planned, with wide streets. Even before the Goldfields' Water Scheme provided it with water, an immense tank with a capacity of one and a quarter million gallons was built a mile and a quarter from town. Southern Cross is built on the western edge of a long, narrow, salt lake.

In 1892, three mines were producing gold—the Central, the South Mines, and the Frazers. The latter is remembered more for one of its engineers than for its gold. The engineer, a well-mannered gentleman named Deeming, murdered his wife in Southern Cross and cemented her body under the hearth at his home.

A happier event in the history of Southern Cross was the first horse race which was held during Arthur Bayley's visit to the town in October 1892. A match was arranged between his chestnut stallion, Ben, and Frank Gerald's bay War-cry mare, Walferd. A man named Henderson acted as a starter and Warden Finnerty filled the position of judge. The amount of money Bayley put on his horse made Ben the favourite, but Gerald's mare won easily.

Although Southern Cross became an active mining town, it was early felt that the gold output did not justify the large expenditure on building the railway. Yet, the Yilgard field produced the largest yield of gold in the Colony for 1891 — 12,833 ounces out of a total of 30,311 ounces. These returns encouraged the general belief that large quantities of gold existed in Western Australia. Experienced miners, after examining the area, were convinced that instead of being the centre of a goldfield, Yilgarn was only on the fringe, and that somewhere in that arid country, stretching to the South Australian border, greater discoveries would be made.

They were right, of course, and gradually the importance of Southern Cross declined. Although the goldfield did not live up to its expectations in 1910, very rich gold, discovered at a shallow depth twenty-three miles north-east of the town, revived Southern Cross for a few years. The town of Bullfinch was built there and is connected to Southern Cross by rail.

The Bullfinch Mine, which has been worked in latter years by Great Western, a subsidiary of Western Mining Pty Ltd, closed down in 1963, thus ending gold-mining of any substance near Southern Cross. As the gold in the different mines was depleted, another industry developed in the district, and has kept Southern Cross going as a centre. This was wheat-farming and it has been carried on with some success, although in a low rainfall area.

Another industry, the mining of iron-ore at Koolyanobbing, about twenty miles from Southern Cross, began in recent years. The ore is crushed, then railed to the Kwinana Steel Works.

However, there is a chance that Southern Cross could have a gold mine again. On 20 January 1971, a report by Jenson Mining and Investment Ltd said that while drilling for nickel, it had intersected five gold bearing zones on the company's Burbridge prospect, near Southern Cross. The directors of the company said that results confirmed the possibility of developing substantial tonnages of low-grade gold reserves. In one hole, sixty feet of massive sulphides containing gold, assaying just over three penny-weights, was encountered below 409 feet. Due to the nature of the sulphide host rock, it was possible that a sulphide concentrate could be saleable as a source of sulphur.

If a gold mine is developed from this new find, and if commercial nickel ore-bodies are discovered around Southern Cross, the town could be restored to some of its former importance. But, as with Coolgardie, its glamour lies mainly in its past.

10
LOOKING BACK

MANY OF THE OLD PERSONALITIES of the goldfields—now scattered far and wide, or retired to live in the city—talk nostalgically of their days on the 'fields. In Perth, special functions are held every year. The old rallying cries—like "Back to Kalgoorlie," "Back to Gwalia," or "Back to" other goldfield towns—are heard again. The reminiscing begins.

"Remember when we were kids and used to climb down the open cut at Mount Charlotte and look for pigeons?"

"Remember those nights out at the old 'Westy Block,' the parties going on till all hours?"

"Those were the days!"

"And did you hear they caught up with old Jim at last and he's doing six months for gold-stealing?"

"What rotten luck! The gold's in the ground for everyone, not just for those blokes in London."

Recalling old times has a cathartic effect on these retired miners and they go home the happier for that.

With the passing of time, events leave an aura of glamour on them. Still, the goldfield does have a special feeling all of its own, as anyone who has lived, say, in the farming districts and then moved to the goldfields, will testify. Whether this difference is to their liking or not is a personal matter, but it does exist.

The people are rather different too. The early arrivals were made up of adventurers, mainly from other states and countries. They had to be physically tough, and very adaptable. Most of them were consumed by the lust for gold, and personal comfort was ignored in its search. Gold often became an obsession and some men forsook all else and spent the rest of their lives searching, going from one 'field to another, occasionally making a strike, but more often than not just getting out just enough gold to live on. Some were lucky and struck it rich and went back to their "old countries" to live out the rest of their lives as gentlemen, but most of them stayed on, and won and lost fortunes from the ground.

These adventurers were mainly "T'othersiders," men who came from the New South Wales and Victorian goldfields and the copper mines in South Australia. Previously these men, or their parents, had come to Australia in the gold-rush and after gold had been found at Bathurst in Victoria. They came from all countries and a lot were miners from Wales and Cornwall.

Both of my grandfathers were "Cousin Jacks." One was a young Cornish tin miner who arrived in Australia to work in the copper mines at Kapunda in South Australia. When Hannan found gold, and started what was to become the richest mile in the world, my grandfather temporarily left a wife and six children behind and came West to try and make his fortune. He never became rich, but did become a "Cap'n," an underground mine manager. These Cornish miners brought their knowledge of mining with them and were an asset in the mining communities. The Americans later arrived with their more modern methods.

My grandmother, a Cousin Jenny herself, followed shortly after with her family. She travelled as far as Albany by boat, then by coach to her first white-washed, hessian home near the mines at Boulder. These women who followed their men to the mining fields had to put up with conditions undreamed of today. It was through their influence that the towns soon became civilised. The goldfields pioneer women cannot be spoken of too highly.

At one time, Boulder had late closing for the shops on Saturday nights. My mother often described how as a girl she always looked forward to them. There was a carnival-like atmosphere as it seemed that the whole town turned out to walk and talk in the streets. The younger ones usually promenaded along Burt Street, turned into Lane Street, around into Piesse Street (which was another street of shops), then back once more into Burt, the main street. The Salvation Army band as well as other brass bands were much in evidence on Saturday nights, so it was the most important night of the week. Sunday, of course, was a day of rest. There was no cooking, and the best clothes were worn. It was expected of everyone to attend church.

World famous celebrities often visited Kalgoorlie and Boulder to appear in their Town Halls, as they still do today. In 1908 Dame Nellie Melba travelled to Boulder to put on a concert. Before it began, she went out onto the balcony overlooking Burt Street, and sang "Home Sweet Home" to a street packed with hushed miners.

A year later Ernest Toy, a world famous violinist and a cousin of my paternal grandfather, gave a recital in Kalgoorlie as part of his Australian tour. His relations and friends were among the large audience which was enchanted by the magic of his violin. Ernest Toy was born in Bendigo in 1880. He studied the violin in Queensland under Charles Manby, and then under Professor Gustave Hallaender in Berlin.

Except for two or three well known incidents, there was little violence on the goldfields. However, several murders were committed and one crime stayed in my grandmother's mind all her life. A few years after she had settled in Boulder, she was walking down the street when she met a neighbour. She greeted him, "Good morning, Mr Smith. How is your wife today?"

"I've just snapped her neck like a bloody carrot," was

his startling reply. He had too, and was later convicted of murder.

My paternal grandfather came out from Cornwall to Bendigo as a young boy with his parents. Like his father, he became a mining engineer, and he and my grandmother came to Kalgoorlie in about 1900. They lived and worked in most eastern and north-eastern goldfields towns following the gold finds. I believe he rode all over the outback on a bicycle, in his search of gold. Cycling was a very popular form of transport at the turn of the century. An ideal path to ride on was the track made by camels. Their flat pads, different from horses' hooves, left a firm, flat path.

For some time, my grandfather was a mining engineer for Claude De Bernales, who was called, among other things, "the handsomest man in the State." Claude De Bernales was one of the controversial figures of the day. Some called him a rogue, but he did give the goldfields a boost at a time when it was badly needed. He arrived in Coolgardie as a mine machinery salesman in 1897, and stayed to become one of the leading figures in the State. After buying some abandoned mines, he floated dozens of English companies. He owned the Foundry in Kalgoorlie and into this he put all the machinery and equipment he had bought with the mines. When one of these mines reopened, the company had to buy back from him the machinery that was needed.

In London, in 1926, he successfully floated his biggest mine, the Wiluna Gold Corporation. The mine was valued at twelve million dollars on the Stock Exchange. By 1936, he became Managing Director of Great Boulder Pty, the richest company in Kalgoorlie's history. Other companies in Claude De Bernales' group provided this mine with its supplies, thus giving him more profit.

De Bernales' financial empire was short lived. In 1939 his companies started to collapse. He was accused of corruption and was forced to resign from Great Boulder. Detectives were sent to investigate by the British Board of Trade, but no charge was laid.

His greatest benefit to the goldfields was attracting four million dollars of English money by floating companies.

The unique, Tudor-style London Court, in central Perth, was built with his company's money. He also built, at Cottesloe, his palatial mansion set in beautiful grounds, complete with imposing statues. It is now the Cottesloe Civic Centre. De Bernales went back to London, where he died in 1963, at the ripe old age of eighty-seven.

Another figure who appeared on the scene around Kalgoorlie towards the end of the century was young Herbert Hoover, later to become President of the United States of America. He was one of the few goldfield's engineers who seriously studied geology, and he was willing to go out into the desert on a camel or by horse and cart, often in intense heat, to examine new discoveries for his employers, Berwick Moreing. His greatest success was the part he played in examining the Sons of Gwalia Mine and recommending its purchase by his employers. He managed the mine for seven months, then left for an important engineering job with Chinese Engineering and Mining Company.

His memorial in Kalgoorlie is a huge Moreton Bay Fig tree in Victoria Park which bears his name: "The Hoover Tree."

In Arthur Reid's book *Those Were the Days*, it is claimed that years after he left the goldfields, Hoover wrote some soulful and passionate verses dedicated to a former Kalgoorlie barmaid. Personally, I doubt this because he was a teetotaller, so would be unlikely to be in the company of a barmaid. He was a very unsentimental type of person and not given to wasting words; and when his appointment in China was announced, and he knew his income would be raised, he sent the following brief cable to his Quaker sweetheart in America: "Will you marry me?"

The goldfields have had more than their share of colourful and interesting characters. Two of these men, while not having the business ability of Claude De Bernales, nor the dedication of Herbert Hoover, nevertheless added a dash of colour to Kalgoorlie.

One was Charlie Nalty, a keen cyclist, who won the six hundred dollar Westral Wheel Race at Coolgardie in 1902, and lived in the reflected glory of the event for the rest of his life. This race was one of the world's biggest cycling events of the time.

He arrived on the goldfields in 1896, and although cycling was his main interest he also loved children, and was generally to be seen where children gathered, always with his beloved bicycle. If it was left parked for a time, he would chain and padlock a wheel to the nearest post or fence. Charlie would often wait outside the school gate with a hessian bag slung over the handlebar of his bike. When the children came out, he handed them carrots or apples, and would willingly tell them of his riding exploits—if any would stay to listen. It was also his habit to wait outside church on Sunday School Prize-giving Days to present his own special prizes to the children—lolly penny sticks.

On Saturday afternoons he generally rode to the Victoria Park, chained his cycle safely to a fence, and rang a large hand bell to call the children into the park. When he had a group of youngsters together he put on a one-man show, telling them stories, playing either the accordion or penny whistle, and teaching them songs. His favourites were "Don't Break My Heart, It's Been Broken Before" and "Irish Eyes." Then he organised races and gave paper-covered toffees to the winners.

Even when he was quite old, Charlie still joined the children on toy train rides at our annual Community Fair and added to the noise and gaiety by vigorously ringing his bell. In the mid-1960s he made his last public appearance at the Fair when he rode a lap round the arena on his bike in the Grand Parade. Over the years he travelled back and forth to Adelaide, where he had business interests, and it was there that he died in July 1970 at the age of eighty-eight.

The other interesting character was an internationally known pickpocket, called "Jack the Jew." His name at times should have been Robin Hood because that was who he often copied. He did his best to equalise the money position in the town by robbing from the rich and giving to the poor. His idea of the rich were the ones who had little better to do with their money than to drink and gamble it away. He would help it disappear a bit faster, that was all. If he knew of anyone who was having a struggle to rear a family or was in some financial trouble, he would often sidle up to them and whisper, "Do you need any money?" It worried him when many who were in need rejected his help, so he helped them anonymously. During the Depression years, he helped many a family by paying their grocery bills.

I met a woman in Perth recently, and she said, "You come from Kalgoorlie, don't you. Do you know Jack the Jew?" Then she told me a story of how she had been left to rear three young children on her own, and was having quite a struggle to make ends meet. One day she found a grocery parcel on her front door step, and every week another one would appear. She said, "I could never see who left it, so one day I waited and watched and saw a young boy leave the parcel at the door. When I asked him where it came from, he just shrugged and said that a man had given him a couple of shillings, and told him to deliver it to my house." Later she found out that her benefactor was Jack the Jew.

Jack would never take a penny from the pocket of a friend. However, if he saw that a man was throwing his money around carelessly while having too much to drink, he thought it was his duty to remove some, and save it to hand back next day.

The police would not allow him to patronise the race-course, which I think rather hurt his pride. In about 1927, he wanted to see the Empire Games, so with only a few pounds in his pocket he set off, and by using his special skills, including card playing, he toured the world.

As he got older, and I think partly reformed, he was often shown off by his friends to visitors. They bet that Jack could remove their watches and jewellery without their being aware of it. The locals always won.

He ended his days in the Government Hospital in Kalgoorlie where he died in 1967. Since then the legends have begun and some stories have grown out of all proportion. My memory of him is on a Mother's Day, when we were new to the neighbourhood with our large family. There was a knock on the door, and outside stood our neighbour, Jack the Jew, with a big bunch of dahlias and chrysanthemums, to wish me a "Happy Mother's Day."

Gold stealing, or dealing in gold, was never in Jack's line, but it was once big business on the Golden Mile. Secreting gold when one was working for a mining company had no stigma in a mining community. A thief who stole anything else would be looked down upon, but not for taking gold. Even today if a man gets caught for gold stealing, he receives sympathy or perhaps indifference, but not condemnation.

A man who worked underground in the 1930s told me how he and other miners stole from the diggings. "I took the middle out of my thermos flask and filled it with small pieces of rich stone. We only got one dollar, forty cents a day wages then and had to do something to get enough to live decently on." I asked him if they were searched when they came up from underground. "No, we weren't searched, but our homes often were if it was suspected we were taking gold. I used to get rid of mine before I went home, by selling it over the counter at the —— Hotel. Later, when the gold detection staff was formed here, and they had a couple of honest blokes who couldn't be bribed, we had to go to the hotel's office to sell the gold, instead of over the front bar." When I asked him how much they were paid, he said, "Oh I used to get four dollars a thermos full, but the fellow who bought it would make a lot more than that out of it. He had to take the risk of getting the stuff treated by putting it in crushings with ore from an almost worthless "show," or have it treated illicitly out in the bush somewhere."

The man who treated it, or the one whose crushings were "salted" with the stolen stone, was sometimes caught, but the big fellow rarely got into trouble. Sometimes, those who had their crushings salted were caught when telluride was found because this showed that the stone came from deep underground on the Golden Mile, and not from the little "shows" in the bush.

He then went on to tell me, "The manager of one mine

thought of a plan to get at least something back from all the gold everyone knew was being stolen from the mines. He proposed to a Board meeting that they have a building on the surface where the men could sell their gold after they came up from underground. In this way the company would at least make something out of it. The Board members did not think much of that idea."

Now that the gold mines are in their dying stages, gold stealing is dying out too. I doubt if we will ever hear about nickel thieves, as it would be a problem for them to know what to do with either the stolen ore or the soot-like nickel concentrate. Thus another chapter in the history of the goldfields is drawing to a close.

While on the subject of the shady side of the goldfields, I must mention another of our local institutions—the Two-up School. When a friend heard I was writing a book, he said, "Don't forget to write about the Two-up. Hop in the car and I'll take you out, and we'll drive past it." I was rather dubious, as I knew it was a strict No-woman's Land and had heard of at least one incident where angry Two-up players had shaken their fists at prying women. However, it was all in a good cause, so out we went, while I slid lower in the car, so as not to be too noticeable.

It was a week-day afternoon and there were about sixty cars parked near a large roofed area with parts of the sides open, so we could see a group of men standing around inside as we drove past. I remarked that there seemed a lot of men there for a week-day, and my friend replied, "Most of them don't work, and this is how they spend their time—in gambling. Of course others play too. Some are tourists who call in, so they can say they have played Two-up at Kalgoorlie."

He added, "When you write, don't forget to say that the game is very well conducted." One of the men who ran it was once a heavyweight boxer, who acted as a deterrent to anyone who might be tempted to start trouble. I was told that one day, when some men were playing the game at a location closer to Kalgoorlie than it is now, they were held up by a gunman. He demanded that they put their hands up

70

and the ring-keeper hand over the money. Unknown to the hold-up man, one of the players had gone into the bush nearby for reasons best known to himself, and, on returning saw what was happening. He crept up behind the would-be robber, and hit him over the head with something solid. After regaining consciousness, the thief was taken to the police station where he was charged.

There have been several different sites for the Two-up School and many and devious were the ways to let people know where the next "game" was to be held. An old rusting bath-heater, lying innocently on the side of the road, was one of the things used to point the way through the bush. However, the National Game has been held on its present site for a few years now and there are at least two unwritten laws it adheres to. The game is never played on the days the local race club holds its meetings, and, to please the housewives, never on the mine pay-days.

Another game in Kalgoorlie which was the rage in 1929, around the time we came back from the Eastern States, was midget golf. It was very popular at the time, and my father dreamed up a similar type of game, and ran it in the Cremourne Theatre. He called it "Pottem" and one half of the hall was marked off in little greens for the game, while his orchestra played for dances which were held in the other half of the large hall. Although I was very young, I remember a marathon dance being held there too, with the last three couples left on the floor in the competition wearily moving their feet, and only just moving them, in time to the music.

A reminder of those days appeared recently in a *Sunday Times* article. Abe Walters, who had gained fame as an orchestra leader in England under the name of Don Carlos, asked for someone to verify his claim for a world record of 108 hours of non-stop piano playing. A friend of my father, and a member of his orchestra, Phil Fryer, came forward and verified Walters' claim that as a fifteen year old boy he had played the piano non-stop for 108 hours. He also remembered that the promoter vanished with Abe Walters' fee. Fryer said, "I went to Kalgoorlie in 1929. We had a dance band going there. I remember Abe playing for 108 hours in the old Cremourne Theatre in the top end of Hannan Street. We used to go along and play with Abe in the afternoons and evenings."

The marathons were common during the Depression. In those days it was simply a case of doing anything to earn some money. The "Pottem" game did not last long as there was too much competition from the more popular midget golf.

Lytton's Theatre, which presented its plays in a huge tent, arrived in Kalgoorlie around that time, and was indirectly the reason for my family leaving town to live in Gwalia. The Theatre hired my father's orchestra to provide background music for its plays in Kalgoorlie, such as *Charlie's Aunt*, *East Lynne*, and *The White Sister*. When the company went on tour to the Northern Goldfields, the orchestra went with them and while in Gwalia, my father found he could obtain a constant job and a house there. So when the contract with Lytton's Theatre expired, we packed up, and left Kalgoorlie for the second time.

I can understand the nostalgia which the old-timers feel for the goldfields. I was born in a house at the top end of Egan Street, Kalgoorlie, and with the exception of six early years, I have spent all my life in the goldfields. From Kalgoorlie we moved to Gwalia, Spargoville, Menzies, the old ghost towns of Mertondale and Vivien, Agnew, Lawlers, Wiluna, Big Bell, and then back to Kalgoorlie again. After seeing the beginning of the end of Agnew, the slow death of Wiluna, and the more rapid one of Big Bell, it is a new and exciting experience to see some of these towns taking on new life again. Some old-timers who have hung on in these places waiting for their towns to "come again" feel the same way, but none thought a mineral different from gold would be the cause of this resurrection. Although they are pleased to see all the activity going on around them, the old gold prospectors shake their heads and say, "It will never be as good as the old days again. There is nothing like gold."

11
SCOTIA

SCOTIA WAS THE NAME of the place we now know as Ireland, the home of the original Scots. This name was also given to a little railway siding on the Kalgoorlie-Leonora line between the old mining towns of Bardoc and Canegrass. Although a lot of prospecting was once done in the area, leaving potholes and dumps behind to tell the tale, nothing of great interest was found—until the year 1968.

To begin the story of Scotia, the third commercial nickel strike, we have to go back to 1962 when Charles Cecil Barton Jones, commonly known as Barton Jones, was running a fence line on his property north of Randalls and parallel to the Trans-Australian Railway line. He was keeping an eye open for indications of copper, which was in demand at that time, when he saw a shed of bright emerald green rock, showing up on the side of a black iron-stone hill. He took a sample of the green stone to the School of Mines for assay but was disappointed to find that it contained no copper. The colour was caused by traces of nickel in the stone.

After realising the importance of nickel when Western

Mining began developing its nickel mine at Kambalda, Barton Jones remembered the green stone, and took another sample in for assay. It contained one per cent nickel. He then took out a temporary reserve over the area.

This was the beginning of the Jones family's interest in nickel. Barton's father, an accountant, came West in the 1880s lured by stories of rich gold strikes. He opened stores in new mining towns, and brought his bride, a wonderful helpmate, to Bulong, a gold-mining town twenty miles east of Kalgoorlie. He became Mayor, and owned one of the two general stores in the town. At one time, his store was burnt down, and had to be rebuilt. The couple reared a family of three daughters and two sons, one of whom was Barton.

The Queen Margaret Mine, the "big" mine of the district, eventually ran out of gold and closed down. The men of Bulong banded together, worked the Great Eastern Mine and kept the town going a while longer. But the town eventually died, and soon there were no customers for the general stores.

In its hey-day, the people of the Municipality of Bulong bought a grand piano for their local hall. Miss Corey, now Mrs Howard, played the piano at all social functions from the age of fifteen. "The Hospital Ball was the social event of the year at Bulong," she remembered with pleasure, "when we often hired Toby O'Toole's Band from Kalgoorlie to play for us and I was able to dance for once, instead of playing the piano all night. The band played the Royal Irish so well that it set everyone's feet tapping, and they just had to dance."

When the town died, the grand piano gathered dust and cobwebs in the Bulong Hall. Old Mr Jones, who had taken up a pastoral property at nearby Hampton Hill, was made trustee of the piano. Occasionally a carload of young people from Kalgoorlie or Boulder broke into the hall, and had a party, using the piano to liven the proceedings.

One day a Roads Board's truck arrived to take the piano back to Kalgoorlie, but Mr Jones, who took his trusteeship seriously, refused to let it go as he said it belonged to

Bulong. However, while he was away on holiday, the piano was removed and taken to the Shire Offices in Kalgoorlie. Mr Jones at last consented to hand his trusteeship over to the Municipality of Kalgoorlie, and the piano was finally shifted down to the Town Hall where it is today. It still remains the property of Bulong and if the town ever "comes again," it is to be handed back.

The property at Hampton Hill meant a life of hard work and real pioneering for the Jones family. A big herd of goats, as well as sheep, were run on the station which was often looked after only by Mrs Jones and the girls, while Mr Jones, and later Barton, went prospecting.

After Barton had finished his education at Guilford Grammar School, where he in turn sent his four sons, he returned to Hampton Hill and in the early 1930s took over the property. He married and he and his wife, Grace, raised a family of seven children, four sons and three daughters. The daughters, Eileen, Lucy, and Mary, all became trained nurses. Two adjoining properties, Coowarna Downs and Coonana, were bought and the four sons, Bart, Bob, John, and Burchell, all helped to run the huge combined property of more than a million acres.

Over the years, Barton worked hard to rear and educate his children and his main ambition was for security for his family. He took on various jobs like ore-carting to help with his finances. He also went prospecting and at one time successfully operated the Blue Quartz Mine. Later he even added wood-carting to his list of money-making ideas.

In 1964, during a period of drought which caused a serious water shortage, the Jones family was faced with the great expense and inconvenience of carting water. At the same time, geologists were hunting for suitable locations to bore for water on the property. This was how John, one of Barton Jones' sons, first became interested in geology. He learned by asking questions. Until then he hardly knew one rock type from another.

An American company became interested in the area pegged for nickel by the Jones family on their property, and took out an option over it. In their negotiations with the company, the family decided to appoint one person as the head of their mineral interest and to act as their spokesman. Their logical choice was John as he was extremely keen and obsessed with the idea of finding a nickel mine. Barton thought it was the job for a young man, and gave John all the time off that could be spared to go prospecting. John learned all he could from geologist friends, and, in common with so many others, found the School of Mines very helpful in identifying rock types for him.

By June 1966, a family conference decided that John should be free to prospect full time, which was no light decision as their numbers were already reduced by Burchell having to leave for two years National Service training. But the Jones family, a closely knit group, always worked together for the benefit of the family.

Occasionally John was called back to help at the station in an emergency, but otherwise he spent all his time prospecting and learning all he could about nickel. His constant companion was his sheep dog, Trelawney, who, when not riding in the back of the Landrover, sat in front of John on the petrol tank of his motor bike.

The world's giant mining companies were beginning their exploration, and already Australian Selection had pegged large areas of Hampton Hill Station itself, even to where the homestead stood. This meant that John had to prospect further afield and to help in his search, he sent to Canberra for a map which showed areas in the Eastern Goldfields which had magnetic anomalies. He had learnt by now the various ways nickel occurred in these anomalies. An aerial map was also sent for and John spent many hours studying these and evolving a theory.

In the area south of Boulder, very many geologists came looking for another Kambalda, but John ruled a line leading north from Kambalda and was confident that his theory should lead him in this direction to his nickel mine. He chose three suitable areas, one at Scotia and two in the Mount Jewell district.

He spent eight weeks in the Scotia district after a year of painstaking work. By his theory, he should have seen some of the signs he was looking for—some gossans or serpentinites or even the blackbutts, the trees that can indicate nickel close by. But he saw none of these and returned home disillusioned. That night he couldn't sleep, so got up and checked his maps again and found that there had been a discrepancy in one which had taken him two miles in the wrong direction.

Taking an assistant, he tried again next morning and recognised the trees and the country he had been looking for, and began pegging claims. He sent a message to his family over the two-way radio, asking them to help with the pegging. It was not long before he found the main indication he had been looking for—an outcrop of gossans on a low ridge. He took samples from this and gave them to his father to take into the School of Mines for assay. Barton Jones, who called himself general rouse-about by now, did all the odd jobs of the place, including a daily trip into Kalgoorlie for messages and to take his grandchildren, by Bob and Bart, to school.

When they received the results of the assays, the readings were 0.415, 0.54, 2.15, 2.7, and 4.15 per cent nickel. John said, "I was astounded, and asked myself over and over again how there could be so much nickel in surface samples."

Secrecy was essential. Too much talk could have brought others around to peg the areas he wanted to investigate. However, rumours soon began to fly and several companies showed interest. The Jones family wanted to have a nickel mine of their own, so for the next few months they worked hard pegging 7,500 acres and sampling, digging costeans (trenches) and drilling. Bart's wife stayed up night after night typing the 2,000 sheets it took for all the claim forms. By September they realised that the job was too big and expensive for them to handle alone any longer, so they reluctantly gave up the idea of running their own mine, and called for tenders.

Barton Jones spoke to Bob Ince, one of the directors of the North Kalgurli gold mine, and mentioned that his son John had pegged good claims at Scotia. Bob Ince asked the

North Kalgurli if it was interested in acquiring the leases, but the company had enough problems of its own at that stage, and declined.

Later Mr Edgar Elvey, the chairman of the Great Boulder, as well as of the North Kalgurli, raised the subject of the leases at a meeting, and the boards decided that it would be beneficial for the two companies to join forces for added strength. They then applied for an option over John Jones' leases. Other world companies were negotiating at the same time, but a twelve months' option was eventually taken out by the partners, Great Boulder Gold Mines Ltd and the North Kalgurli (1912) Ltd. Barton Jones, the old gold prospector, was especially pleased that the local companies were the successful tender.

The General Manager of the Great Boulder, Mr Mitchell, was the partner's negotiator. He was of Cornish mining stock and as a matter of interest, his father was the manager of the Burbanks Mine at the time of the big robbery. Money from the sale of a large quantity of gold was stolen by thieves who got away with their haul and were never apprehended.

The Great Boulder's interest in nickel began soon after it was discovered at Kambalda. The company's first venture was with the North Kalgurli when they jointly applied for a temporary reserve at Lake Rebecca. In this, they were unsuccessful as it was granted to other companies.

Mitchell, who had recently taken over the management of the Great Boulder, was also searching for other gold mines. The Great Boulder Gold Mine was beginning to phase out and needed a large low-grade deposit, similar to the Mount Charlotte operation run by Western Mining in Kalgoorlie, or a small rich deposit to help keep the mine's treatment plant working. He became interested in the old mine at Mount Martin, south of Boulder, and took out three twenty-four acre leases for gold over the area. This was on Hampton Properties, part of Hampton Plains Estate Ltd. The plains were named after Governor Hampton by Lefroy during an expedition to look for pastoral country in 1863. This large landholding, first floated in 1900 as the Hampton Plains Estate Ltd, was later included in the properties of the Hampton Uruguay Ltd, a large London syndicate. It comprised 189,192 acres of freehold land with full mineral rights—the only land-holding of the kind in Western Australia—and 292,932 acres of land held under pastoral leases from the Government of Western Australia. Coolgardie just missed being included in the syndicate's holding.

The Mount Martin Mine still had several small ore-bodies so the Great Boulder began exploration work on these. When nickel became big news, Mitchell went out to Mount Martin and recognised ultra-basics in the old mullock bins at the mine. He took samples from the bins for analysis and found they contained 0.5 to 1 per cent nickel. Negotiations then began for an option over the Mount Martin area and took twelve months to accomplish. A survey over the whole area began and at the time of writing, the assessment of its nickel content is still being undertaken by the Great Boulder.

As soon as the option agreement for Scotia was signed in May 1967, the mining partners began exploration work. The drilling programme got off to a disappointing start. Number One Scotia drill was lost, so they began again seven feet away. Within six weeks, the drill discovered rich nickel sulphide. Assays showed they had found the richest nickel in the world at between 800 and 900 feet down. The sharemarket responded favourably to this news.

By August 1968, the companies announced that they had a mine and the rest of the negotiations went through. The Jones family received cash, shares and royalties on all nickel produced from the mine, and the security so desired by Barton for his family was assured.

Surprisingly, the family's way of life changed very little. Work at the station went on as usual but John bought a new Landrover and continued prospecting till the time came when the family formed their own company, Jones Mining N.L. Barton Jones and his wife, Grace, realised a life's dream and set off on a world cruise in February 1970.

The nickel mine at Scotia, forty-five miles from

Kalgoorlie, was developed six miles in from the railway line. A journey on that line, although painfully slow, was always interesting, at least to me when I was a child. In spring, the driver and fireman picked a spot to stop the train and "boil the billy" where a carpet of everlasting and wild flowers spread out over the countryside. This gave the passengers a chance to get out and stretch their legs and to admire or pick the flowers. Often there was no need to wait for the stop. The old steam train, pulling its dog-box carriages behind, with water-bags swinging outside each compartment door, travelled so slowly on upgrades that one could step off, gather a bunch of flowers, and catch up to climb on the train again while it was still chugging uphill. Stores and mail for the few prospectors in the district were left at the little siding of Scotia.

Work at the Scotia nickel mine began soon after the option was exercised. A main shaft was sunk and a poppet-head built. Offices, machinery, sheds, men's living quarters, and a canteen soon followed. By late February 1970 the companies were ready to begin carting ore for treatment into Fimiston at the Great Boulder plant.

At midnight on 8 December 1969 the Great Boulder Gold Mines Ltd crushed its last gold-bearing ore. A major producer for over seventy years, it had paid out about twenty-two million dollars to its shareholders in dividends. On 12 December, four days after its last crushing, a group of twenty-six people watched a gold ingot of 400 ounces poured at its treatment plant. This marked the end of the company's gold operation and the beginning of its nickel era. Gold, which had reigned supreme since 1893 on the Golden Mile, was beginning to topple from its throne. Its place was taken by a base metal.

Although the North Kalgurli (1912) Ltd is still working, they have finished developing the mine. When ore reserves and gold-bearing columns are treated, this mine will also close and its plant converted for treating nickel, possibly by another company.

While ore at Scotia stockpiled, the Great Boulder's treatment plant was being converted to become the second nickel processing plant on the goldfields. An ore-train terminal, with bridge-bin and feeders, was built, as well as storage facilities and nickel concentrate conveyors. Milling of the ore began on March 5 1970, and the two ore-trains delivered 2,500 short tons per week to the plant at Fimiston, the suburb in which the Golden Mile is situated. In the joint companies' quarterly report on 17 October 1970, it stated that 2,786.5 short dry tons of nickel concentrates were railed to Western Mining, which buys all Scotia's concentrates. The percentage in concentrates was nickel, 12.69, copper, 0.66, and cobalt, 0.20. It also said that in an earlier shipment, the concentrates contained platinum in commercial quantities.

A survey was taken among the miners at Scotia to gauge whether they preferred to live in a town built near the mine, or to commute daily from their homes in Kalgoorlie and Boulder. The majority chose the latter, so one large and three small buses were used to transport the men free of charge to and from the Twin Towns for the three shifts worked at the mine. About thirty men live from Monday till Friday in the single men's quarters provided at the mine, but most of them spend their week-ends in town. The only females living in Scotia in October 1970 were a cook and two helpers who had their own quarters.

All water needed at Scotia is carted by road in special trucks, because except for salt water struck while sinking Scotia's main shaft, there is no other water in the dry, scrub-covered district.

Twenty-two miles further east is another joint venture by the North Kalgurli and Great Boulder. This is Carr-Boyd Rocks which shares Scotia's amenities and bus service. Although the mine has a large work-force, shaft-sinking and developing, the only people actually living at Carr-Boyd Rocks are the caretakers. This nickel strike has a high copper content and will be treated separately from Scotia's ore. To help with the water problem, a small desalination plant has been installed at Carr-Boyd Rocks to produce a supply of fresh water from water pumped out of the shaft.

The flamboyant teller of tall tales, explorer and prospector, William Carr-Boyd, had a group of rocks named after him when he discovered them during an expedition in the late 1880s, and now his name is further perpetuated by a nickel mine found nearby. Carr-Boyd, at one time a reporter for the *Bulletin*, writing under the name of "Potchossles," was much travelled and a fine bushman. Although Ernest Giles was the first white man to visit the Warburton, it was Carr-Boyd who put it on the map. When he reached Coolgardie in 1896, he reported splendid looking auriferous country in the Barrow and Warburton Ranges. He also told the unlikely story that it had not rained in that area in the twenty years since Giles explored there and that he had clearly seen Giles' footprints.

"Smiler" Hales, editor of the Coolgardie *Review*, wrote of him thus: "When not out in the wastes of the 'Never-Never' Carr was at home to the public in one of the local halls. He hummed tunes on a gum leaf, told blood-curdling stories of his life, and took up a collection. At one of these 'At Homes' Carr-Boyd threw a desert scene on a magic lantern sheet. 'There I am,' said Carr, 'almost at death's door through want of a drink.' 'Carr,' said a voice, 'who took that picture?' 'I took it myself,' said Carr."

His tales of gold finds fired the imagination of his listeners and were partly to blame for the ill-fated Kimberly Rush. He was one of the few men of his time who had sympathy for the Aborigines. This paid off when at one time a native saved him from dying of thirst.

He explored many miles of the inland waterless country in Western Australia and was confident he would find payable gold in the Cosmo-Newberry Hills (which is now considered an interesting area for nickel exploration). Unfortunately all he found was low-grade ore, which, after the cost of carting it long distances for treatment, would not have been a paying proposition. In spite of his years of prospecting and exploring, Carr-Boyd failed to make a fortune and he finally went back to Victoria to settle down with his family and tell tales of his exploits in the "Wild West."

Movie-goers in Australia and overseas have been able to see some of our "Wild West" in Western Australia's first full-length film, *Nickel Queen*, which was partly filmed at Broad Arrow, twenty miles from Scotia. Some other scenes were shot at Ora Banda and in the bush country nearby. The film created a lot of interest in the district and shootings of the scenes were all well attended by crowds of onlookers. In several scenes the old Broad Arrow pub was used as the home of Meg Blake, the Nickel Queen, played by Googie Withers. The old iron and wooden building was one of the town's original hotels and has altered very little in the years since it was built. Broad Arrow and Ora Banda were perfect settings for the film without the need for building artificial sets. The ghost town of Ora Banda was used for several outdoor shots and the dilapidated buildings of the town, including the old stone hotel with its ant-eaten flooring and flapping iron roof, were just what the film men wanted.

We went along one Saturday afternoon to see some of the filming at Broad Arrow. The setting was the railway station, so we sat down on a pile of railway sleepers over the track to watch. The stone walls of the station looked as solid as the day the last station master left in 1925, although the doors, windows, and part of the roof, were not so well preserved.

As we watched re-take after re-take of a scene, I thought it would not take long for the glamour to go out of film making. John McCallum was a most exacting film director.

Nickel Queen has a special interest for us on the goldfields as many local residents were used as extras. This, combined with the authentic background, gives a realistic goldfield's flavour to the film—except perhaps for the appearance of the hippies, as we haven't quite developed that far yet!

The heavy volume of traffic going through Broad Arrow reminded me of my journey north in the previous winter. I made a quick visit to some of the places I had known in my younger days. Now these towns are coming alive again through nickel exploration.

12
A VISIT TO
THE NORTHERN
NICKEL-FIELDS

THE SIGNPOSTS AT the corner of Maritana Street in Kalgoorlie point to the top end of Piccadilly Street and read — Leonora 146, Menzies 82, Laverton 219, Sandstone 255, and Wiluna 336 miles. We left on this road in a utility heavily laden with new tyres, drill rods, and groceries, bound for a diamond drilling crew which was working forty miles north of Agnew. My driver was Charlie Gianni, who has lived on the goldfields all of his life.

As we bumped across the railway line leading to Leonora at the top end of Piccadilly Street, I glanced over to where Simpson Newland once lived in his battered old camp. A dear old gentleman and a great friend of children, he wanted to live to be a hundred and receive a telegram from the Queen, but couldn't quite make it. He died six weeks before his hundredth birthday.

We drove on through a stretch of ground scarred and pitted with pot-holes and dumps, the scene of a great deal of prospecting and dry-blowing, till we passed the Tea Gardens on our right. A few old-timers still live there, just outside the town boundary so they don't have to pay

TOP: Gianni syndicate with a rich "show" near Ora Banda. Crushings average three ounces to the ton. BOTTOM: Carting nickel ore from the Otter Mine, Kambalda

WABCO

BLACK SWAN HOTEL
ALEXANDER'S
UP TO DATE
GREAT SALE
CLEARING SALE

rates or conform to building laws. Some of their buildings are masterpieces of contrivance.

There are not many left living there now because as soon as one of the occupants dies his camp is removed, as the authorities feel that the Tea Gardens are an eyesore. Those remaining are mainly bushmen who are happier living this way, rather than in conventional homes. I asked Charlie if he knew why it was called the Tea Gardens and he told me that in the horse and buggy days, there really was a tea garden for travellers using this road.

Later we passed Paddington, or the place where it had once been; there lots of mining had been done in the past. When we were about seven miles from Broad Arrow and crossing a dry lake bed, Charlie said that in 1946 heavy rains filled all the surrounding lakes and covered the road up to a depth of eight feet. For weeks, till the water subsided, row boats were used to take food and essentials across to the people of Broad Arrow and beyond. The first essential item was, of course, beer. In Broad Arrow itself, one man was nearly drowned when he was elected to swim from one side of the road to the other to get a boat. Sixteen inches of rain, about six more than the yearly average, was recorded in the town during that time.

There is little left at Broad Arrow now and the only gold mining done is perhaps the odd "show" being worked around the district. One prospector near the town wears his shoes back to front. He finds it easier to keep his balance that way as he has lost the bottom part of both legs. He camps alone and dollies and pans off his stone in a sitting position. When he needs stores, or some company he must battle his way into Broad Arrow, a mile away.

Broad Arrow was once quite a big town and, in its heyday, a municipality. It had a large hospital with a special fever ward, two breweries, eight hotels, a Resident Magistrate, several stores, drapers, blacksmiths, a dramatic society, and even boasted a Stock Exchange.

The town's name was derived in an unusual way. Among the first to leave Kalgoorlie for the new find in 1893 were O'Mara, Quinn and Pyke, with their pack horses. O'Mara

Top: Hannan Street, Kalgoorlie, taken in 1904. Bottom: The Mt Charlotte Reservoir, taken at the opening ceremony in 1903

told his nephew, who was to follow later, that he would leave arrow signs on the trail for him to follow. J. E. Tregurtha later recounted: "My mate and I, following on a few hours later, saw these marks on the left hand side of the trail. Each was from two to three feet in length. Jim Read, a well known prospector, was the first I heard suggest to call it The Arrow."

The first three leases were pegged late in 1893 and one, the Golden Arrow, worked in conjunction with several other leases worked by the Golden Arrow Mines Ltd, produced 19,645 ounces of fine gold from 34,727 tons during the years from 1893 to 1906.

I asked Charlie why there were so many dumps, all close together, near the town. He explained that they were earth taken from shafts sunk into the deep leads. They followed the course of ancient river beds which were buried when the earth folded over and covered them. These sunken rivers are good sources of gold nuggets and a large number of the population of Broad Arrow was engaged in this type of mining: tracing river beds from twenty to fifty feet down, and from one to about five feet in width. Now all the area around the dumps is pegged for nickel.

Charlie told me that fourteen miles from here, at Ora Banda, his family and he had worked, as a syndicate, a fairly rich gold mine called the New Mexico. Crushings from this mine averaged three ounces to the ton, and they worked the mine for several years after the last World War.

A few miles from Broad Arrow a big ore-train passed us filled with nickel-ore from Scotia and bound for the Fimiston treatment plant.

We travelled on past the remains of the old towns Bardoc, Goongarrie, and Comet Vale. The latter was named after a comet had been seen overhead at the time. Bardoc, about twelve miles from Broad Arrow, is mainly remembered for the "Bardoc Murder." J. Tobin and a mate, while working on a claim near the town, noticed some digging had been done on a shaft which had apparently been abandoned and filled in. Curiosity made Tobin dig down into the newly filled-in hole, and to his horror he discovered a fully clothed body with the head smashed in, presumably with a pick. This was before the turn of the century and although an investigation had been made, the death remained a mystery.

I have been told that it is unwise and not in good taste to name one's home with any Aboriginal word, beginning with "Goon" or "Goonya." Most nurses treating native patients in the goldfields know what it means but I wonder if the person who named Goongarrie understood the meaning. The mining town once had two hotels and two stores but nothing is left now. There is a story told by Mrs Walshaw, a daughter of Dan Baker who operated the condensers at Goongarrie and was credited with the discovery of Comet Vale. Mrs Walshaw said, "Fred Breton kept the hotel on the main road. A man called 'Wild Cattle' from Queensland, a good horseman and as wild as they come, bet his mates one day he would ride his horse into the hotel bar. This he did in fine style and called for drinks all round. Suddenly, the pine flooring of the bar-room gave way and the horse and rider fell through into the cellar where an old black gin was washing bottles. Several bottles were broken and the gin was killed. A meeting of diggers was later called and a verdict of accidental death recorded."

Further on, we passed a group of shafts belonging to a mine called the "Young Dago," six miles from Menzies, then on into Menzies itself. The town was named after the discoverer of the field. It still has a store, hotel, and a stone Town Hall and Shire Offices. Menzies has been a centre for pastoralists, and, these days, for the nickel men. In the past, the big mines were the "Lady Shenton" and the "First Hit."

While we waited to fill our vehicle with petrol, I looked over towards a mine where many years before our family stayed for a few weeks at the manager's house. I was twelve years old at the time and although I forget the reason for our being there, I do remember that I was very fascinated with our host. One day I peeped into his study and saw it was filled with murder mystery books which overflowed onto his table. And when I heard that he slept with a revolver under his pillow, it completed the picture I had formed of a very sinister character. After that, I was always very subdued in his presence. In actual fact he was just a quiet man, preoccupied with the problem of running a gold mine.

His wife, who was very large, had a passion for hats and hooked rugs. The house was so stuffed with furniture that one had to squeeze through one's way about the house.

My sister and I attended school for our short stay in the town and our teacher's name was Mr Rule. Our father used to quip, "Every inch a teacher, every foot a rule." The rule I remember most was the one he hit the children with. As we left Menzies, I glanced over beyond the railway line and saw the old State School. It looked just as it had been thirty-five years before.

There was little to break the monotony between Menzies and Leonora because the road does not go through Kookynie or the stony waste of ground which is all that is left of the once thriving town of Malcolm. Kookynie still has a store and hotel as well as the ruins of several stone buildings.

We came in behind Mount Leonora and pulled up in the main street of Leonora for a meal at a café. The town has changed little in the thirty-odd years since I last went there. The three hotels which were serving customers in those days are still standing but only two are now licensed. When we lived there, Bert Webb ran the "White House," Peter Hill the "Central," and Mrs Crammeri the "Commercial" hotels. Over the road, the very old and time-worn "Exchange" Hotel still stands, although it was never used as a hotel in my time but only as a boarding-house. The painted sign, BED AND BREAKFAST 2/6, still shows clearly on the front wall.

Bearded men were much in evidence in the café and hotels and all seemed busy talking about the one subject—nickel. Discussion ranged from likely areas, and who'd pegged where, to the latest share prices.

We had left Kalgoorlie around midday, so by the time we drove away from Leonora and left the bitumen behind, the day was wearing on and we were heading straight into the afternoon sun. All the way up, we could see every now and then coloured plastic streamers tied to trees as markers. Charlie told me that all the ground near the road was pegged for nickel and associated minerals.

The ride was not quite so comfortable now we were off the bitumen and on the dirt road and had the sun in our eyes. About forty miles from Leonora, and half way to Agnew, we reached Doyle's Well. The well was still there but the hotel had gone. A cracked, empty and weatherworn cement swimming pool was next to the place where it once stood. Faint whispers of an old scandal of a nude swimming party held there in the moonlight came back as I gazed at the pool.

Mount Clifford is about four miles to the east of Doyle's Well. Western Mining holds leases there and has announced a nickel strike. As we travelled along near there, we could see I.P. (Induced Polarization) lines marching away in straight lines through the bush. The country around this area and further north is very bare except for dry-looking shrubs and trees, some killed by the drought. It was eerie to listen to the stark, dead limbs rattling in the wind. There had been some rain which had turned stretches of the ground near creek beds into what looked like bright green stains. Unless more rain fell, this short grass would soon turn brown like the rest of the countryside.

Once or twice after leaving Kalgoorlie Charlie jerked his head towards the bush saying, "Good country that." I looked but couldn't see what was good about it, then he added, "Good nickel country." He explained that he thought the lay of the country looked good as there were some hills near a salt lake. "Notice how many of the nickel strikes are near a salt lake with hills nearby?" He added, "All this ground is pegged though."

It was dark when we went through Lawlers and all we could see was a single light burning in the only building left standing, the old police station, now owned by a nickel company working in the area.

Finally we reached the hotel at Agnew, about six miles further away, where I was to spend the night while Charlie went on to take his cargo to the drill rig, forty miles further on. He would come back to pick me up next day. I had not been back to Agnew since I left there in 1942 and had been looking forward to seeing it again for very special reasons. It was the town where I was married and where we built our first home. There was no hotel at the time I lived there, but a row of shops, with additions, has since been turned into the Agnew Hotel and is now run by an old friend, Mrs Trundle.

Parts of the old Lawlers Hotel which had been dismantled were used to help build the hotel at Agnew. As I entered the front bar, I recognised the top half of a tall, dark, mirrored sideboard which once graced the dining-room of the Lawlers Hotel. The bottom half, as well as a solid, carved, wooden trolley, also from the same hotel, were in the hotel's dining-room. On the dining-room wall hung a wooden-cased clock which once kept time at the Emu Gold Mine office at Agnew.

There was a handful of men in the bar. The main topic of conversation was, of course, nickel. I asked one man what activities were going on around the place and was soon given a run-down of all the work being done by companies and individuals within a radius of about a hundred miles. It is harder to keep secrets in scattered communities than in the bigger towns.

Early the next morning Bill Cox kindly drove me around the town to show me what was left of Agnew. The first stop was where our home once stood. Except for the brick and cement foundation for the kitchen stove, there was nothing left to see. Next we drove to the old mine, a part of which is still being worked on a very small scale. The Emu Gold Mine shut down practically overnight when Claude De Bernales' companies folded up. The water has come up over the old levels underground but men who have worked there say there is still plenty of gold underneath. A lot of these old mines could come to life again if there was a big enough rise in the world price of gold.

As we drove back to the hotel, memories flooded into my mind, but all there was left to see were cleared spots and some signs where homes or camps once stood. Rusting sheets of old iron and a heap of smashed bottles marked the site of the shack where the "Broken Hill" sold "sly grog" to its customers.

Unfortunately there was no time to drive the five or six miles out to Vivien, although in any case there probably would have been little to see. It was a ghost town at the time we lived there before the second World War, when my father treated the sands from the old mine by the cyanide treatment method. In those days there was nothing but our temporary home and the cyanide treatment plant. I remember that our bough shed near the kitchen was often visited by a four-foot long bungarra, which left footprints in the sand like babies' hand marks. With its tongue flicking in and out, it always headed for the Coolgardie cooler where it knew eggs were kept, till someone noticed it and chased it away.

My mother always felt that there should be some old sovereigns which had been lost in the old town site. In fact we heard a story that there had been a hoard of them buried at, or near, the Vivien Gem Mine two or three miles away. We did go searching there sometimes but found nothing. Once, after a rain, my mother suggested that we look around for sovereigns, and I actually walked over the first one she found. It was really only a half-sovereign. We found it resting on a shilling piece near a cellar where, I believe, a Mr Coyne once had a shop. A few months later, my mother had a bigger find near some old tent pegs. The rain had washed some soil away and she saw something glinting in the sun. She picked up a stick and dug, and found not one, but six half-sovereigns!

Charlie arrived back at about ten o'clock. He had delivered his supplies, so good-byes were said, and we set off on the homeward journey. Now I could see Lawlers in the daylight. We passed the spot on the hill where the old Lawlers Hospital had once been situated. I never knew the hospital but I came to know a private house which had been converted from the hospital. This private building, now gone forever, had the reputation of being haunted. Twice I visited friends who lived there and stayed overnight. The first time I was given a bed in the room with my friend's children. I woke in the night to hear the sash window opening and closing, opening and closing, but I couldn't see anyone near the window and the children were asleep. During my second and last visit, I slept in the lounge room as I refused to stay in the children's bedroom again. The room was dimly lit by the still glowing embers of a fire when once again I woke suddenly during the night. This time I heard heavy breathing, although I could see that I was alone. My head went under the blankets till morning. Afterwards, whenever I was invited to stay there overnight, I always politely refused. Now that the old hospital has been pulled down, I expect the "ghost" has disappeared too.

At the foot of the hill, the shell of the old Lawlers Hotel still stands. Although Lawlers was once the biggest town on the East Murchison, at the time we lived in Agnew all that the town consisted of was the hotel, a hall, the police station, post office, and two shops. Later the shops moved their business to Agnew as it progressed.

At one time the State battery was working there, managed by my brother-in-law. His house and two or three others were scattered through this once big town. On the way to the battery, we passed the remains of the Daisy Queen mine, south-east of the town. The house, which once was the manager's residence belonging to the Great Eastern Gold Mine, was occupied in those days by the MacPherson family who were treating the sands belonging to the mine. The house had once been a proud home, and a large one too, with six bedrooms.

There were at least six hotels in Lawlers' hey-day, but the Lawlers Hotel which I remember was owned by Mr Wertheimer, and was always known to us as "Werthy's Pub." To the people living in Agnew, town really meant Werthy's Pub, as it was our social centre.

Through the middle of the building ran a wide passage, called for some reason "The Cremorne," which was used mainly by us young folk as a dance floor. The central courtyard of the Big Bell Hotel was always referred to as the "killing pen," and I was told it was so named as the place where the men took the ladies for a drink and then proceeded to "slay them." But I never did find out the reason for naming the passage in Werthy's Pub "The Cremorne."

Tables, chairs, and a sofa or two were arranged down each side of the Cremorne, but pride of place was reserved for the pianola. This has since been donated to a convent. In its old age, it has retired to a place of respectability which should somewhat atone for the wild life it had lived for many years in Werthy's Pub.

As long as there were two or three couples and someone willing to pedal the pianola, any time was dance time for us. "Shine On Harvest Moon" and "Lily of Laguna" were only two of the many rolls we danced to. Sometimes my father pedalled it for us while we danced or gathered around to sing.

There was a "drunks' room" at the back of the hotel where guests who were not considered respectable, or sober enough, were given lodging. It was nearly always occupied, often by a man in from the bush on a "bender."

But now the corrugated iron hall in Lawlers has gone too. We held our Balls there which went on till the small hours. Formal attire was worn on these occasions as it was our only chance to really "dress up" in long frocks and the men in their good suits. One of the best bands which played for us in this hall was the Slav's Band, consisting of guitars and the big bass — or "burdar" as its owner called it. I can still hear its deep "thrum-thrum" giving the beat, so our feet could not help but keep in time. During the night, the "Alberts" were always put on by popular request. This was a chance to show off our full skirts as they flew out when "Corners swing" was called.

Unfortunately there was not time for too many memories as Charlie was anxious to move on. So with a last look around, we drove off.

We lunched at Leonora, then headed for Gwalia, two miles away. An electric tram service once ran between the two towns, but now there is no sign of rails to show where the track had been.

A new caravan park on the edge of the town was filled with caravans mainly belonging to people connected with the nickel search. Further on, near Half Way Creek, we saw a mobile van and a shed used for testing rock and soil samples for those prospecting for nickel.

All around Leonora and Gwalia are the dumps from old shafts and mines, but none of these, except the Sons of Gwalia Mine, lasted for any great length of time, although the Trump, the Forrest, Gold Blocks, and the Tower Hill did well for a few years.

An alluvial gold find, discovered by Booden in 1896, was the reason for establishing the town of Leonora. After his camp had been destroyed by some of the Aborigines, Booden went to Cue and brought back his mate, Sullivan, who pegged land which later became the township. A few months later a prospecting team, Carlson, Smith, and Glendinning, financed by Tobias, a Welsh storekeeper at Coolgardie, found the reef which was the beginning of the Sons of Gwalia (the old name for Wales) Mine. This mine produced over two and a half million ounces of gold for a profit of four and a half million dollars till it closed down on 28 December 1963. The main shaft was sunk at a steep angle underlay, instead of the more widely used vertical shaft.

Horses were used for about forty years to pull the ore trucks. Some of the horses lived out most of their lives in the darkness underground. When they were finally replaced by mechanical methods, their eyes were blindfolded before they were brought to the surface and light was only allowed gradually to their eyes which had been used to darkness, or the limited light from carbide lamps.

Big cockroaches, up to two inches long, bred in the underground stables. To relieve the tedium of their shift, the miners often staged cockroach races at "crib" time and placed bets on their favourites.

Miners brought home their waterbags filled with condensed water, still steaming hot from the condensers. When the water cooled, it tasted better than the bore water used in the town, especially towards the end of the summer when the wells ran low and the water seemed thick with magnesia.

Although I was only a child when I lived there, the memory of the odour from the men's underground clothes comes back quite strongly. Their working clothes, flannels and dungarees, grey with dirt, always seemed damp and had a pungent smell of a mixture of carbide, explosives, dirt, and sweat.

We drove around the town to see what was left of Gwalia. When the mine closed down and the miners left, the businesses and schools closed, leaving almost a ghost town behind. Some homes were left standing but others were dismantled. Mostly only rubble was left behind: strips of hessian, twisted corrugated iron, wire, and old timber.

The State Hotel, the one State-owned hotel that made a profit, still stands on the corner looking towards the railway station, but now it is empty and vandals have thrown stones through the windows. On the other side of the street there is a shamble of timber supports and flapping pieces of old iron; all that is left of Major's old boarding house. During the last war, the building was taken over by evacuees from Perth at the time when the air attacks on Broome caused fear that Perth could be the next to be bombed. These evacuees were mainly wives and children of doctors and professional men. They named their new home, "La Hacienda." Now the mine staff houses on the hill look badly in need of repair and the swimming pool nearby gapes open, cracked and empty. Surprisingly, we could see that most of the houses still standing are occupied and when we called in to visit Harry Hoffman, he told us there are over a hundred people living in Gwalia. These are mainly the overflow due to the lack of accommodation in Leonora, which could not cope with all of the extra people who have arrived in the town in connection with nickel exploration in the district.

Harry Hoffman's humble little home is near the empty playground of my favourite seat of learning, the Gwalia State School, although after the Gwalia mine closed down, the school was transported elsewhere. Harry told us that he had only paid twelve dollars for his home and had originally intended it to be his base camp, but as things had turned out, he found he was spending quite a bit of time there.

Harry is the man who realised the commercial potential of the Weebo Stone. A gold prospector of many years, he saw some pretty stone with mauve and yellow markings while on one of his fossicking trips about sixty miles north of Leonora. The stone is unique. It has a design of concentric squares, with each portion a different colour or shade.

A building firm became interested in buying the decorative stones to use in making feature walls, and were willing to pay $488 a ton for it. Seeing that there were approximately 50,000 to 60,000 tons in the deposit, Harry Hoffman thought his fortune was made. He borrowed two thousand dollars to develop the site and had mined thirty tons of stone when his second application for permission to mine the stone was refused. Earlier, in March 1969, a Mining Warden had over-ruled objections by the Native Welfare Department and made a conditional recommendation that Mr Hoffman's application should be approved.

However, the area, said to be important to the Aborigines, was subsequently declared a sacred site.

The case of the Weebo Stone, so called as the deposit is on Weebo Station, twelve miles from the homestead, attracted a great deal of publicity and controversy at the time. Harry Hoffman even had his life threatened and a university student in Perth pegged the War Memorial in King's Park as a protest against the Weebo pegging.

In October 1970, a year after the controversy, the story was revived when a report was made that the thirty tons of stone already mined had been stolen from the site. Hoffman thought that most of the stone was stolen before the area was placed under a ministerial reserve. He put it bluntly thus: "It stinks. I am the loser and the natives don't give a damn about the stone anyway." It is believed that a now disbanded tribe visited the site regularly and that older tribesmen regarded the stone as sacred, but there have been no known pilgrimages there in recent years.

Hoffman said that after the publicity over the stones had died down, some older Aborigines offered to exchange some of the stone for bottles of wine. He had even seen some of their women sitting on piles of stone even though one legend has it that any woman touching the stone will disappear.

Harry has a slab of polished stone leaning against the wall in his house and he asked if I could guess what it was to be used for. Not being good at guessing games, I confessed I couldn't. So he told me: "That is to be my gravestone." In the meantime, it stands there as a symbol of his buried hopes of becoming rich.

He was at a loose end and not working at the time we met him and he explained, "Before the interest in nickel, a man could go out and work a 'show' till the gold cut out, then move on to another likely place. Now all this country is taken up by the big companies, and it has pushed the gold prospector out." But Harry Hoffman is not the type of man to stay down long, and in spite of setbacks he has bounced back again. His interest in life and sense of humour have seen to that.

Having come to know many old prospectors, I have never ceased to be amazed by their philosophies. Most I have met are nothing like the general picture that is painted of them — men living in the bush, usually under very rough conditions, with only one thought in their minds; the mineral they are looking for. Though few have had any more than a basic education, they have time to think and read and they make some surprising observations.

One said to me, "What legacy are we leaving for those coming behind us in say, a hundred or two hundred years' time? Pollution alone is a big enough problem, but do you realise we are denuding the earth of its minerals? See that transistor radio? It contains about two and a half pounds of copper. Each big jet needs about seven tons of nickel and look at all the tin cans we are using and throwing away. The world is using its minerals at an increased pace and not putting anything else back, and there are only so many deposits we can draw from."

We visited an old prospector and his son who were living in a camp which can only be described as functional for living. It was very enlightening. In a bookcase made of grubby boxes stacked on top of one another were piles of well thumbed books, including the war memoirs of Winston Churchill and the complete works of Shakespeare and Dickens. The old fellow said to me, while puffing his pipe, "I just love Dickens' women. There are no others like them." The younger man quoted reams of Shakespeare to a generally unappreciative audience when he was quenching his thirst at one of the hotels in town.

Even old Jack Cairns, one of the last of the dry-blowers, living in his group of shacks south of Coolgardie, writes ballads. I have found that some of the old prospectors can be very interesting people.

Charlie and I finished our conversation with Harry Hoffman and set out for the last part of our trip home, arriving just before dark. I had a strong feeling of having just returned from a journey to the past.

13
POSEIDON

POSEIDON, GOD OF ALL THE SEAS in Greek Mythology, and proclaimed pagan god in 1969, now rules a dry kingdom. Because Mount Windarra, near Laverton, is about as far from the sea as can be imagined.

The prospector who discovered the Poseidon leases at Mount Windarra was Ken Shirley, whose parents lived at Leonora at the beginning of this century. Herbert James Shirley, Ken's father, owned donkey teams and used them for carting general goods from Mount Malcolm (the rail head) to the outlying towns of Leonora and Gwalia. He continued this service until the railway was extended to these towns; he also made trips to Darlot, Mertondale, Lawlers and as far north as Wiluna.

Herbert Shirley sold out his carting business and took his wife, three sons and two daughters closer to Perth. While they were living in Queen's Park, a Perth suburb, their son Ken was born. Not long after, the family moved again, this time to Adelaide where Ken spent his schooldays. At the age of fourteen he went with his family to Coober Pedy and the work on the opal fields aroused his life-long interest in mining. Opals did not bring the high prices then that they do today, and although the family unearthed some beautiful opals the Shirleys did not make their fortunes.

The news that rich gold had been found at the Granites, 350 miles north-west of Alice Springs, reached Coober Pedy and inspired young Ken Shirley and a mate to join in the rush and try their luck. It meant that they would have to cross over hundreds of miles inland, mainly desert country, so they obtained five camels to carry them and their provisions. Horses were ruled out because they require watering twice daily. On the other hand, camels can manage quite a few days without water. This was an important consideration, as the prospectors faced long stretches with an uncertain water supply.

The camels only travelled twenty-odd miles a day, so the journey was long and arduous. Half-way to their destination they met the rush in reverse. A lot of disappointed miners were returning from the Granites, because the gold, though rich, had not lasted long. After speaking to some of these men, and realising there was no point in going on, Ken Shirley and his mate decided to try their luck on the Arltunga and Winnelkes fields, fifty to sixty miles east of Alice Springs. These fields had been discovered many years previously, even before the time when gold was discovered on the Murchison in Western Australia.

Ken Shirley worked on these fields for eighteen months, till the gold became harder to find, then transferred his attention to the Tennant Creek goldfield which had just begun. It was here that he spent the greater part of his life, from about 1934 to 1963, with the exception of the war years which he spent serving in the R.A.A.F. He mined on his own behalf, and was married in Tennant Creek after the war.

When gold became scarce, he was employed by exploration companies, still working in the Tennant Creek area. Later Mr and Mrs Shirley moved to Queensland where they found a beautiful spot on the coast, north of Cairns. Here they decided to settle and build their home. Ken Shirley had decided to give up his mining life, so he bought and ran

TOP LEFT: John Morgan, one of the partners to bring in the first nickel samples from Kambalda to Western Mining Corporation. TOP RIGHT: Ken Shirley, prospector, who discovered leases at Mt Windarra for Poseidon. BOTTOM: Jack Lunnon, the first driller to intersect nickel sulphide at Kambalda

a hotel. He found that this new way of life was not at all to
his liking, so when one of the mining companies at Tennant
Creek offered him the chance to prospect for them in the
Cox's Find area in Western Australia, he jumped at the idea.
This would also enable him to see his relatives again, as,
although by this time his father had died, his mother, a
brother, and sister were living in Kalgoorlie. After the hotel
had been disposed of, Ken Shirley and his wife came West.

The three-month contract of employment extended to
six months, at the end of which the couple decided to return
to their home in Queensland. However, a chance meeting
with an old friend, Clem Wegener, a geologist, changed
these plans when Shirley was offered a job with a mining
company. He accepted and worked around the Kalgoorlie
district for six months. This was followed by work with
another exploration company whose interests were in the
Laverton area. Now fate stepped in and took a hand. He
said, "If they had told me I was to work around any other
place, I'd have turned the offer down. But Laverton—that
was different."

This lonely little town marked on the maps at the edge of
the desert had long captured his imagination. He had read
and heard of most Central Australian expeditions either
beginning or ending their journey there. Men with their
camel teams had left this last civilised outpost before
venturing into the wilderness that did everything it could
to claim their lives. Such a man was Harry Domeyer who
led the expeditions to search for Lasseter's Reef, using
Laverton as his stepping-off point. After months of unsuccessful searching and the daily problem of finding adequate
water, he returned to Laverton. Domeyer lived out his last
years at Kalgoorlie.

Laverton is a strange place—the land of Wongi, dry
mulga and spinifex country, of searing heat and bitter cold,
and, in between, glorious spring and autumn days. In
recent years I stayed there for a holiday and fell under its
spell. The charm that part of the country held for me was
an intangible thing. While there I learned about the Wongi
culture, and even attended a female corroboree, and I felt

TOP: View of the Kalgoorlie suburb of Lamington about 1900. Homes were
mainly built of timber, iron, and hessian. BOTTOM: A modern "suburb": single
men's quarters and canteen at Scotia. All units are transportable should the mine
cease to function

in Laverton that I was very close to the heart of our ancient land.

The town was originally known as the "British Flag" after the mine discovered by Tom Potts, Harry Dennis, and George McOmish, all originally from New South Wales. The men showed specimens from their mine at a big exhibition held by the people of Coolgardie at a time when the growing town of Kalgoorlie was taking the glamour from their town. Doctor Charles W. Laver was so impressed with the rich stone from the "British Flag" that he bought an interest in the mine and decided to live there. Subsequently Laver contributed so much to the development of the district that when the town site was selected by Warden Burt, it was named Laverton after the doctor.

In his book *Coolgardie Gold*, written about the time Laverton came into being at the turn of the century, Albert Gaston writes: "The district carried a large population of the most happy-go-lucky, free and easy lot of men I ever met on any part of the goldfields. The mail arrived once a week (by coach) and on Sundays everyone for miles turned up at Skipper's store for their mail and a week's supply of rations, also to take part in, or to watch, the weekly cricket match."

The principal mines working in 1901 were the Niagara, Lancefield, Euro, Craigiemore, Childe Harold, the Ida H., and the Augusta. My grandfather once had a tribute on the Augusta, which he worked successfully for some years. When my father was a boy, he and a friend heard there was to be a corroboree in the bush and decided to watch. The tribes in that part of the country had a bad name in those days, so the boys knew they had better not be seen. That night they crept through the bush, and at a safe distance watched the corroboree. My father said of that night, "We could see the dancers' painted bodies by the light of the fire and on the outskirts some girls were sitting with their heads bent and eyes closed. Suddenly, one quickly lifted her head to have a peep at the forbidden sight. One of the dancers saw her and brought the nulla-nulla he was carrying down on her head with a sickening crash." My father and his friend thought they had seen enough and crept away as

102

silently as their thudding hearts would allow and decided they were no longer interested in corroborees.

Warden Burt, who gave Laverton its name, had a town named after him. It was Burtville, now a ghost town eighteen miles from Laverton. After its discovery in 1900, it was considered to be the richest field for prospectors in the West. This was because its rich reefs were small enough for individual prospectors to work and did not attract the companies.

The town consisted of two hotels, two stores, and two banks. Mail was delivered by bicycle, privately operated, at a charge of ten cents a letter. Martin Walsh was reputed to have declared, "The least thing this rotten government could do for the prospectors is to give 'em a weekly mail once a fortnight."

The town must have had a dark side as one early resident, W. R. Ritchie, wrote: "The town had rather a sinister reputation and up to the time I left, all the people in the cemetery had met with violent deaths, mostly murders and suicides."

Ken Shirley prospected near Laverton for twelve months, pegging several ultra-basic groups for the company which employed him. After this another proposition was put to him, again by his friend Clem Wegener. It resulted in his working for Poseidon.

This mining company had been formed in 1952 for the purpose of mining sheelite and wolfram deposits (both used for hardening steel) at Hatcher's Creek in the Northern Territory.

Ken Shirley had been prospecting in the Laverton district for some time and decided to work along a low, sandy, scrub-covered range about ten miles long. This was about six miles further on from the old Lancefield Mine which at one time was one of the best low-grade gold producers of the back country. Discovered in 1899, it had a fairly long life, although it did close down and re-open a few times, before finally closing in May 1940.

While working on the range, Ken found ultra-basic outcrops and took several samples from the southern end of Mount Windarra, which was part of the range. It was mid-summer on 12 February 1969 when he took these samples into a Kalgoorlie assay office for testing. The results, which he collected on 27 February, were nothing unusual — except for one sample. This showed .48% copper and .71% nickel and caused Ken Shirley great excitement. Not wanting the news of his find to get out, he told no one, and went straight back to Laverton and pegged the area the next day.

Then began more sampling and pegging and by 10 March 1969, the area at the north end of Mount Windarra, which later excited world-wide interest and caused an upheaval on the stock exchanges, was pegged. In all, Ken Shirley pegged thirty-seven claims for Poseidon, covering about fourteen miles of serpentinite with banded iron contact. These leases also contained a number of gossans, most of which assayed 1% nickel, or better, distributed within five miles. He was very happy with his dealings with the company and said the arrangements under which he was contracted to Poseidon were very satisfactory to both sides.

A firm of consulting geologists, Burrill and Associates, was hired to take charge of the work of estimating the quantity and quality of the ore reserves. The population of Laverton, about fifty in all, were well aware as time went by that something big was going on near their town, so several of the townspeople bought shares in Poseidon. Rumours spread to Leonora, only seventy-three miles away, and some lucky people bought Poseidon shares which at that time were selling at between eighty cents and one dollar each.

Late in September 1969 there was apparently a leakage of information which sent Poseidon shares moving up rapidly. The information was confirmed on 1 October when assay results of 3.5% nickel and 0.55% copper in the Number 2 hole on the Shirley zone (named after the prospector) was announced.

The Stock Exchanges were ready for a new strike to start them moving again, but no one could imagine the effect a few assay results would create, or that it would begin

Australia's new mineral boom. Poseidon was the match which lit the rocket that sent the mining market soaring.

Within a few days, thousands of acres surrounding the Poseidon leases were pegged and registered. Then most of the men, apart from company prospectors who had pegged the claims, had nothing to do but wait till the offers came along and to option or sell to the highest bidders, and there were plenty of bidders. It is estimated that millions of dollars were made by pegging around Poseidon.

Suddenly, little Laverton was famous all over the world. Few people who read the newspapers had not become familiar with its name. Nearby Leonora woke from its sleep when it found itself in nickel country, and joined Laverton in playing host to visitors from everywhere. V.I.Ps, newspaper reporters, television crews, mining men, and peggers converged on the towns and their total of three hotels just weren't enough to accommodate all the people.

A flurry of excitement was caused when a major shareholder, engineer Norman Shierlaw, nicknamed Mr Poseidon, flew up to Laverton to look over the leases. The town's airstrip was being used daily instead of weekly by mail planes or the Flying Doctor. English T.V. star, Peter Wyngarde, flew up to have a look and said of the drill rigs dotted over Mount Windarra, "Marvellous, fabulous, just so exciting." World famous novelist, Hammond Innes, also flew in to look at Leonora, Laverton, and the Poseidon leases as part of his visit to the nickel country while gathering material for his next book.

Out at Poseidon's leases at Windarra, drillers and field hands were busy both at their jobs and trying to keep visitors away. KEEP OUT signs, hammered on trees or posts were used as a deterrent. If these were ignored, the visitors were politely asked to leave. As Poseidon's shares went up and up, and more interested people wanted to see what was happening out there at the leases, exaggerated stories got around about guards waving shotguns at them.

Unwelcome visitors could be kept away from the ground, but little could be done about the aerial spies, and there were plenty of those flying overhead every day.

The graded, dirt road leading from Leonora to Laverton did not take long to become covered with about six inches of fine, red dust caused by the heavy traffic and was a source of great annoyance to people travelling over it. In places near the road there are signs of the old railway line which once branched off from Malcolm and led to Laverton. It also serviced the mining towns of Murrin-Murrin and Mount Morgan.

When Mount Morgan townsite was ready to be pegged in 1898, the mine whistle blew, announcing the time for the pegging to start, and the race was on to peg the best blocks. An overweight publican and an ex-runner, racing to peg the same block, caused much amusement to the onlookers. Much overpegging was done, and in fact one block was pegged five times over. Finally a Warden's court was held in Malcolm to sort the mess out, but only two of the five contenders turned up in court. The magistrate decided to settle the dispute by telling the rivals that whoever reached the block and got his pegs in first would be the lucky man. They were several miles from the block in question so they dashed out of court, jumped on their horses and galloped off. The winner rode a racehorse he owned, which easily streaked away and won from his rival's ordinary old nag.

There was little for visitors to see at the Poseidon leases except for some percussion and diamond drills working, although an experienced eye could learn something from the colour of the dirt thrown up by the percussion, and of the sludge from the diamond drills. Featured in the 21 December 1969 issue of the *Sunday Times* was a photo of the first drill hole at Poseidon's leases. It was not exactly the hole itself, because that was covered by a box with a hole cut in the seat, and surrounded by a rough corrugated iron shelter. The hole had been percussion drilled to a depth of 350 feet, and since it had served its original purpose, it was now put to a more functional use as the nickel company's lavatory.

As I looked at the picture, I was reminded of another bush toilet that existed a few years ago. This belonged to the Doyle's Well hotel and was built over an old mine shaft. The hotel, about forty miles from Leonora on the road to

Wiluna, was really a way-side inn mainly used by people travelling between Leonora and the north country. It was a stopping point for the mail truck, on which I travelled many times. I always felt that I took my life in my hands every time I used that toilet, for I had visions of the sides of the old shaft caving in, and the seat, occupier, and all, falling down into the depths.

The picture of the Poseidon "A" hole also appeared in the English *Daily Mirror* under the heading, "Worth a fortune—the mine that looks as if it's only worth a penny." The story said, "As anyone in the city can tell you, the Poseidon nickel shares are the talk of the market. They are rocketing as if the nickel down there were pure gold. This picture, however, may bring a few city gents down to earth a bit. It was sent by cable from Australia yesterday—the day the wonderful stock climbed above the $107 mark on the stock exchange before settling back to $104.

"Here, in black and white, is what the mad scramble is all about. In the centre of the picture is a test borehole shielded by a corrugated iron cover that looks a little shaky. The whole structure looks disconcertingly like a certain sort of outhouse, but the resemblance is entirely accidental."

This article had no effect of quietening the stock market. On 9 December a buyer made history by buying one share on the Perth Stock Exchange for the world record price of one hundred dollars, and some felt that this would be the limit. They were proved wrong. During February 1970 Poseidon shares reached their record of two hundred and eighty dollars each.

Poseidon's neighbours, members of the Windarra Club as they called themselves, hoped the nickel extended into their claims. Rumours that an ore-body from Poseidon's leases was dipping towards their claims sent shares in their company moving upwards. This happened to several companies, but the shares of a few of them remained relatively high, mainly those considered close enough to Poseidon's ore-bodies to have a chance. Drilling will, in time, tell the true story of these companies' chances of success.

Samin Limited was placed in charge of Poseidon's nickel venture and in June 1970, it advertised for a general manager for the Windarra project. The advertisement stated that the successful applicant would take control of operations involving mechanised underground mining and ore treatment by conventional concentration methods. It was then expected that the manager would control a work force of at least 600. Samin was also considering additional treatment facilities for the Windarra nickel and copper concentrates.

In 1971, the new manager arrived to begin his new duties. By then, the mine was already taking shape. A new exploration shaft had been sunk, and part of one side of Mount Windarra was bulldozed away ready for two declined shafts to burrow down into the nickel ore-body. Plans were also under way for a new town for the expected work force and their families.

Also in February 1971 Poseidon became a major shareholder of Lake View and Star Limited gold mine which intends converting its gold treatment plant on the Golden Mile to treat nickel sent down from Mount Windarra. This will be the third major gold mine on the Golden Mile, which intends to convert its plant for treating nickel. The Great Boulder already treats ore from Scotia, and plans are under way for the North Kalgurli's Croesus plant to treat the nickel ore from Selcast's proposed mine at Spargoville.

The biggest headache the manager of Poseidon will have to face will probably be the shortage of men to mine the nickel ore. It is bad enough for the mines starting up near the larger, settled towns to obtain, and hold, their work force, but it will be much more difficult in the remoter areas. Laverton is 219 miles from Kalgoorlie, and even in Kalgoorlie it is hard enough to acquire a work force large enough for the needs of the gold mines of the Golden Mile. There are just not enough trained men, particularly underground machine miners and winder drivers, to fill the requirements of the local gold mines, let alone the new nickel mines.

Many of the gold miners have been lured away by the nickel bonus and the opportunity of making big money. The gold mines have had the problem of recruiting and training a labour force only to lose them to the nickel mines. Married men are recognised as being more stable than single men, but their employment in a mining town entails problems such as proper housing, schools, medical services, and so on. Fortunately for Poseidon, in nearby Laverton there is a kindergarten and a primary school, as well as a hospital visited by the Flying Doctor, and a resident doctor is stationed at Leonora. There is even a branch of the State Library at Laverton.

However, mines which will eventually open up further north, even longer distances from settled towns, will find themselves faced with bigger work force problems. A few ideas have been put forward to overcome the problem of distance. One has been to commute the workers by air from established towns, while another was to provide trains, complete with showers and change room, so that the men could clean up on the way home.

This is only the beginning of the Poseidon story and the rest is in the future. No matter what the rest of it will be, and even after the mine is worked out, the name Poseidon will be remembered and go down as one of the legends of the goldfields.

14
EVERYBODY HAD A SHARE

THE EFFECT OF THE share market boom created by Poseidon will become part of the goldfields' history. It will certainly be remembered by a lot of people for a long time. The boom had a tremendous effect in Australia and the whole world, but in the Eastern Goldfields, near the cause of it all, it would have been hard to find a family who were not affected by it. A type of madness hit the goldfields towns. The word "shares" was on everyone's lips. People who had never in all their lives gambled in the ordinary sense began speculating recklessly on the market.

Rumours were rife. One day it was "Buy so and so shares. They are only twenty-five cents but they should be up to five dollars by the end of the week." The next day another share would be in favour. Someone would know a man who knew someone who knew the brother or sister of a prospector, and the word spread around. "You had better buy XYZ's shares as they are on to something good out there." One woman who did know the sister of a prospector was lucky. She took her friend's word and invested all her small savings, one hundred dollars, in the company's shares and made two thousand dollars from the deal. There were dozens of such cases.

Local interest in the share market began before the Poseidon boom, as Western Mining had earlier made share issues available to its staff members. Later on, the North Kalgurli (1912) Ltd and the Great Boulder Pty followed suit. Members of the staff were able to buy their company's shares at a low figure and many sold to make a handsome profit. Some used this money to travel, others bought new homes, furniture or cars, while still others set themselves up in business. Having had their first experience in dealing with shares and finding it very profitable, the majority of these people went on to play the share market.

For a time, the market moved on the wildest rumours. It was said later that some stories were spread deliberately. One rumour that an exploration company had struck 9% nickel while drilling for water had a rocketing effect on their shares but they came down later when nothing more came of the report. Some people made a lot of money from that incident, but others who came in late lost heavily.

In the few months when everyone was share-crazy, some odd things happened. One Sydney broker told of a syndicate that had printed gold-embossed scrip costing seventy-five cents each. A few hundred thousand were ordered, then someone in the group pointed out that after the cost of the printing, there would be no money left for the mining venture.

In any place where more than one person gathered, the talk naturally turned towards shares and everyone had his "dead certs." A newspaper reporter in Leonora, covering the story of the biggest Warden's Courts in the history of Western Australia, said, "I have never seen anything like it!" He was in the bar of the local hotel when the share report was read out over the radio. Everyone stopped talking and there was complete silence till it was all over.

Share syndicates sprang up everywhere. Businessmen, shop assistants, labourers, housewives, teachers, and even high-school students, all formed syndicates or bought shares singly. The board at the front of the Stock Exchange in Kalgoorlie, next to the Palace Hotel, always had a crowd gathered around, craning their necks to read the latest share prices.

The Steak Bar in the Palace Hotel, where most of the nickel men congregated, was where a great many of the share tips came from. The problem was to pick the right ones from the many being offered. Some men made — and lost — fortunes from hot tips picked up in the Palace Hotel.

What a wealth of history has been made within the walls of this hotel. It was built in 1897 on the site of the first tent hospital in Kalgoorlie. Many world celebrities made it their temporary home because, although there were dozens of hotels in Kalgoorlie and Boulder, it was considered the fashionable place for the V.I.Ps, Royalty, Prime Ministers, Governors-General and stage celebrities to stay

Word of the Palace Hotel's fame as a source of information must have travelled quite a distance as our local newspaper, the *Kalgoorlie Miner*, received a letter from a man in Malaysia, who wrote: "What I cannot obtain is the vital information regarding exploration activities of the mining companies. I am sure there are individuals or firms who supply such information because in the Eastern States there are investment counsellors who inform their sub-scribers by weekly or monthly newsletters.

"To be frank I am a subscriber to such newsletters but because of the great distance I feel handicapped, for by the time I read the news, the subscribers in Australia have already bought and pushed the market up.

"In order to overcome this I want a direct contact.

"By the way, do you know of a Mr Paddy Hannan (either actual or nickname) who, I understand, specialises in reporting from the goldfields for the investment coun-sellors in the Eastern States.

"I quite understand that news of some importance does not occur every day, just once in a while, and it is just such news or rumours that I seek. I am told that the rumours always originate from the Steak Bar at the Palace Hotel in Kalgoorlie, where most of the field-crews meet at the week-end.

"I wonder if this is true and if so, a contact here would be of great value.".

For about six months, the excitement and the money-making went on and the world was a lovely place. The day of reckoning came when the bubble burst. It was not a sudden bursting — more like a slow leak — but it certainly came. The share prices gradually declined, with an occasional rise to make one think the decline was just a temporary measure, but the facts were hard and cold. The fall continued and Poseidon led the downward trend just as it had led the boom. Several people who got out at the right time were lucky, but a great many were caught as the shares tumbled down.

During the boom, the end of the financial year was too far away to think about taxation but as June approached this became another worry. The Taxation Department announced that money made from speculative shares would be added to the shareholder's income and taxed. Several speculators had not thought of taxation, and were caught. Some people faced bankruptcy and at least one person had to sell a business bought through his luck with share transactions. Many were forced to sell out at a lower price than they had paid for their shares, to save going bankrupt, while others, those who could afford to, left their shares hoping that the companies would eventually strike nickel, or oil.

Although there were several factors involved, the decline of the share prices on the Stock Exchange was mainly blamed on the Tasminex affair.

15
THE TASMINEX AFFAIR

THE TASMINEX AFFAIR, as it was called, rates an important place in the nickel story for the part it played in shaking the confidence of the share market, which really never recovered.

Mr Stan Bridgman, ex-mining projects officer with the Native Welfare Department, first became interested in the Mount Venn area in 1967 when a Wongi showed him a small vein of copper there, but it was not economical enough to work. He had a good relationship with the Aborigines and later helped a group of Wongis to peg thirty-three claims at Mount Venn which were optioned to Western Mining.

Mount Venn is located on the Cosmo Newberry native reserve. When the agreement with Western Mining was finalised Bridgman told another group of Wongi prospectors of a spot, also at Mount Venn, where he had found outcropping copper, and sent them to peg it. They pegged seventeen claims which were optioned to Tin Creek Mining Company and explored by Tasminex N.L. who had the right to acquire a sixty per cent interest in the prospects covered by the claims.

114

Martin Watson and his colleagues, Johnny Greenwood and Thomas Murray, formed the prospecting syndicate. Raymond Watson, Martin's brother, had also been a member, but was fatally wounded by a policeman, who was defending the life of another policeman, in an incident on the Laverton native reserve in September 1969. He was buried at Leonora and fifteen months later a group of Aborigines returned to Leonora to perform a symbolic second burial.

Although they had conducted their own ceremony at the same time as the Christian burial, the second ceremony was considered necessary because the Aborigines believe that a person's spirit, his *kuran*, remains in his body till it is captured and, in the case of a young person, transferred to some one else of the same age. According to his relatives, Raymond Watson's *kuran* was very restless and disturbed, and took longer to capture and transfer to the one chosen to receive the spirit and to live in harmony with it.

The ceremony began with the friends and relatives lying face down at their camp, while the not so close relatives walked around the grave, and then retired, wailing loudly. They painted white lines on their bodies, then collected leafy branches which they took back to the camp and threw over the people still lying there. This was the signal for them to stand up and for the ceremony to be over.

The first and second ritual burials are the same, except for the first taking place at sunrise and the other at sunset. Between these ceremonies, the dead person's name is never spoken in case the *kuran* hears his name being called and enters the wrong body.

When the Aboriginal prospecting party, led by Martin Watson, took members of the Tin Creek Mining Company and a geologist to where a copper vein outcropped at the foot of Mount Venn in September 1969, they could not find anything on the surface to guide them to an extension of the vein. Martin Watson searched the surrounding ground minutely and found match-head size grains of copper carbonate. He carefully tracked these to the entrance of a small hole fifty yards away. This, Watson explained, was a mother rabbit's nest where she had borne her young outside of the general warren area. While digging the hole, the rabbit had dug out the grains of copper carbonate where they had become scattered by weather conditions.

Late in January 1970, rumours circulated, first in Laverton and Leonora, then in Kalgoorlie, that while the company was sinking a bore for water, they had encountered mineralisation which proved that it had good nickel values. This bore had been sunk right where the mother rabbit had built her nest. Tasminex's shares began rising rapidly on the rumours.

At this time Bill Singline, Chairman of Tasminex, flew from Melbourne to inspect the leases. On his return on 25 January he was interviewed by journalists and was quoted as saying, "It could be bigger and better than Poseidon. We found massive sulphides and more while digging a cellar near the drill site. We struck them at only ten feet down and that was about 500 yards from the drilling rig."

When asked if any figures were available on the nickel strike, he said, "No, we're a bit in the dark because we have not got onto the geologists, Hall, Ralph and Associates, in Sydney yet."

Following this and similar newspaper reports, Tasminex's shares, which seven weeks before had been only sixty cents, rose to ninety-six dollars before dropping back again after the consulting geologists dissociated themselves from the report a few days later. Poseidon shares were then selling at their peak and the market was looking for a successor to take over from the glamorous leader. After the first rapid upsurge of share prices, it seemed likely that Tasminex could be the one to take over the title, so the journalists, interested to find the reasons behind the sudden turn of events, eagerly awaited Bill Singline's return from his visit to Mount Venn.

Singline, although a successful businessman, appeared to have had little experience with public companies or to be aware of the sensitivity of the share market at the time. He had no idea of the effect his remarks were to cause. On 28 January 1970, a news item in the *Kalgoorlie Miner* said

Top: A ten-head battery near Widgiemooltha in the 1920s. Bottom: The nickel concentrator plant at Kambalda

that Ralph Hall and Associates had written to the Stock Exchange to say they wished to dissociate themselves from all recent newspaper speculation about the merits of Tasminex N.L. nickel prospects. They also said that they "regarded Cosmo Newberry (on which Native Reserve Tasminex's leases are situated) as a routine exploration prospect in a zone of ultra-basics, with associated magnetic anomalies and surface nickel values."

Early in March 1970, the company released a statement which showed only minor nickel content in Mount Venn. The report said that the two holes drilled to date showed traces of nickel in the first bore and low values of nickel and copper in the second bore. After the release of this report, Tasminex shares, which that day had opened at sixteen dollars, twenty cents, closed at seven dollars, fifty cents a share on the Melbourne Stock Exchange. The company's report continued: "The Board of Directors regrets the unwarranted fluctuations on the shares on the Stock Markets and advises shareholders to treat Mount Venn in its proper perspective—an exploration prospect of considerable promise, but as yet unproven."

The report was greeted with hoots of derision in the visitors' gallery and on the trading floor of the Exchange.

On 13 October 1970, a report was tabled in the Victorian Legislative Assembly concerning the Tasminex share dealings. It was made by a Launceston barrister, Mr J. W. Wilson, who was appointed by the Tasmanian government to investigate dealings in Tasminex shares between 7 November 1969 and 18 March 1970. Mr Wilson said that the directors of Tasminex N.L., Tin Creek Mining Corporation Pty Ltd, and the companies' consulting geologists made huge profits by selling Tasminex's shares in late January and early February.

The directors of Tasminex expressed surprise and criticised several of the report's conclusions. This was on 16 October 1970, by which time Tasminex shares had dropped to three dollars, fifty cents.

Many people who bought and sold quickly at the right time made a lot of money from Tasminex shares, but others,

TOP LEFT: A diamond drill rig at Poseidon. TOP RIGHT: The opening of the two declined shafts into the side of Mt Windarra. BOTTOM: The first road train taking nickel ore from Scotia to be processed at Fimiston on the Golden Mile

who had bought at a high figure too late, lost heavily. Some were the local amateur share punters misguided by the "good thing," and some families are still carrying the burden of debt they incurred by buying Tasminex shares. Many of these thought they could get in and out quickly, without having to raise the actual cash.

The Tasminex affair caused the beginning of the end of the gay buying and selling of the speculative shares. Perhaps those exciting days are gone forever, but the optimists are waiting for a new rich nickel strike or the discovery of another glamour metal to revive the market and set it in motion again.

16
THE PEGGING BAN

THE CHANCES OF MAKING a new nickel strike were delayed for a few months because, with the pegging of millions of acres of land for nickel and associated minerals, and the resultant massive paper work, the Mines Department became bogged down. The Minister of Mines, Mr Griffith, announced a halt to pegging on Crown Land on 21 January 1970 and the rule became fully operative thirteen days later on 3 February. Some of the sections of the Mining Act were amended, for the twenty-fifth time since the Act was first introduced in 1904.

During the ban, people continued to peg, although this was against the regulations and their claims could not be lodged. The only areas they were still allowed to peg legally were stock routes, private property, and, with a permit of entry, native reserves. Prospectors searched for likely areas, hoping that if they found anything favourable they would be the first to get their pegs in after the ban was lifted.

Naturally the delay made them feel very frustrated, but it was not until one of the amendments had got as far as the

Legislative Assembly that they were up in arms. The amendment which aroused such strong feelings was the proposal to restrict pegging to a small area of the State, and later to throw open other areas by granting exploration licences for a maximum of one hundred square miles. The small prospector realised that he was going to be pushed out.

While Mr Grayden, M.L.A., and others put up a fight in Perth to have the amendment removed, prospectors, pastoralists, businessmen, and parliamentarians formed an association at Boulder aimed at protecting the interests of people involved in mineral exploration and production in Western Australia. They decided to send a committee to Anzac House in Perth to protest against the amendment.

However, this was unnecessary as two days later, 5 May, the amendment was withdrawn from the Bill. So prospectors, as well as mining and exploration companies, settled down once more to wait and guess the date of the lifting of the pegging ban.

Towards the end of May, after a prospectors' meeting at which the Minister for Mines said that the ban should be lifted within a few weeks, expectation grew to fever-pitch on the nickel-fields. Many part-time prospectors took annual leave from their normal employment in expectation of the end of the ban, while exploration companies stood by with four-wheel drive vehicles loaded with pegs and equipment ready to go. It was said that one company alone had ten vehicles loaded with pegs in readiness.

To everyone's relief, on 29 May it was announced that the embargo would be lifted on Friday 5 June. New regulations, mainly concerned with marking off for pegging, would be given out at the same time in a special printing of the *Government Gazette*. Now, with their period of waiting and inactivity over, the men prepared for the race to begin again. There were fears that disputes would arise if several men wanted to peg the same blocks. At first it was reported that copies of the *Government Gazette* would be posted up outside mines departments after midday on 5 June, and this conjured up visions of hundreds of men shoving and jostling to read the notice first, then a mad scramble for

120

vehicles, and possibly wild and reckless driving to reach their selected areas first.

To overcome this problem, and to make pegging fairer for everyone, it was next announced that marking off instructions would be broadcast over the radio a few minutes after midday on 5 June. This enabled the men to leave town in advance and be at their chosen sites ready to start pegging as soon as they heard the new rules over their radios.

In the last days of waiting, the streets of Kalgoorlie and Boulder were busy with vehicles loading up with petrol drums, stores, water, and, of course, pegs. The offices of Mining Registrars had a steady stream of prospectors, examining maps to see which ground was still available for pegging.

The day before the lifting of the ban, and especially in the morning of the day itself, the sky was noisy with aircraft. Most were flying northwards. However, except for some early activity, the towns were fairly quiet on the day which one newspaper reporter called "Fizzog Friday." There was an air of excitement and impending drama because newspapers had built up reports of possible shootings by men ready to defend their claims with guns.

The front page of the *Daily News* featured a radio-telephone interview with the then Premier of Western Australia, Sir David Brand, who was on board the *Centaur* with Lady Brand on their way home from an official visit to Japan, Hong Kong, and Singapore. He warned against a Wild West reaction and appealed for law and order to be maintained in the "Great Nickel Rush." Sir David said he hoped that "common sense would prevail and there would be no danger to life or limb." Not many local people thought there would be any real trouble but it all added to the general excitement and speculation.

Television cameramen and newspaper reporters were in town on the big day but there was little out of the ordinary for them to report. Most of the prospectors were waiting to peg many miles away. There was little left to peg around Kalgoorlie itself as the ground that had not been previously

claimed was mainly taken up by mining companies'
temporary reserves.

I asked a friend of ours who was going pegging if many
prospectors were taking guns in the bush. He shrugged and
said, "I don't know. Anyway, I have a couple of equalisers in
the car," and showed me two pick-handles. He grinned and
said, "These should equalise me with a young bloke if he
tries anything." However, he went on to say that the handles
were really for his picks and that he doubted if there would
be any real trouble. He was proved right.

After the new requirements for pegging and marking off
mining tenements were announced over the air at about
seven minutes past twelve on 5 June, all the men hurried to
get their pegs in as soon as possible. One problem was how to
obtain the five foot high corner boundary posts. Before the
altered regulations, three foot pegs were used. Most men
solved this problem by tying two three foot pegs together.
The size of the blocks was unchanged, still 300 acres, but
about four times the number of pegs were needed because
extra pegs had to be put in to mark each fifteen chains
between the corner posts. There were several other changes,
but the pegs, or lack of them, were the men's main worries.

A few teams had taken out the longer pegs with them and
this made some people suspect a leakage of information
prior to the lifting of the ban, but perhaps they just guessed
right or remembered that at a prospectors' meeting a few
weeks before in Coolgardie, which Mr Griffith attended,
one of the suggestions was that if five foot pegs were used
they would be seen more easily through the bush.

The peggers went out in all directions from Kalgoorlie,
but a lot of attention was centred around Mount Clifford,
thirty to forty miles north of Leonora, as half of a temporary
reserve held by Western Mining Corporation had been
thrown open for pegging. Temporary reserves held by
mining companies must be divided at the end of every
twelve months and one half relinquished. It was well
known that Western Mining had made a nickel strike at
Mount Clifford and as a result, the areas nearby became a
target for peggers.

Some pegging teams were as well organised as army campaigners, and immediately after they had heard the new regulations they moved to their predetermined positions and began pegging. They used walkie-talkies for communication.

Just two and a half hours after midday on Pegging Friday, the first three claims were lodged at Leonora. Within the next fortnight, hundreds of applications were submitted to the mines departments on the goldfields, the largest number being at the Kalgoorlie and Leonora offices. Now the mines department had another pile of paper work to tackle and the Warden's Courts were faced with more problems of disputed claims and overpegging.

Although the majority of prospectors went north, a lot of pegging was also done to the south of Kalgoorlie. I saw this when I travelled that way not long afterwards.

17
BUS TRIP
TO THE SOUTH

AT EIGHT O'CLOCK on a cold morning I left by bus from the depot at the Kalgoorlie Railway Station. The bus was bound for Esperance but my destination was Norseman, 120 miles away, and the southernmost town of the goldfields. The driver took us down through Boulder, on the road to Kambalda, over Gribble Creek, and past the Boulder cemetery. A glimpse through the bush showed us parts of Hannan's Lake.

In an area near the lake, nine and a half miles south of Kalgoorlie, is the site chosen by Western Mining to build its thirty million dollar nickel smelter. The decision to build the smelter on the goldfields instead of alternative sites will have far-reaching effects on the district. The news was hailed by the people of Kalgoorlie and Boulder as an important step in the development of the area. A significant factor in the selection of the site was the belief that more nickel deposits would be discovered in the Eastern Gold-fields, besides those already found at Kambalda, Scotia, Nepean, Carr-Boyd, Laverton, and Widgiemooltha.

It is necessary to build the smelter outside a six mile

radius of Kalgoorlie because of the Clean Air Act, but there are several other reasons why the Hannan's Lake site was chosen. The smelter needed to be readily accessible to water supplies and serviced by road and rail. Local flora was considered to be fairly resistant and had overcome the effects of sulphur dioxide after the initial damage in the early days of gold mining. The prevailing winds blow mainly from the south-east and south-west and lessen air pollution in the towns. Western Mining gave the assurance that the rate of sulphur dioxide would not exceed the acceptable rate of twenty parts per one hundred million.

The smelter is only part of a two year development programme for the nickelfields, costing $75 million. Western Mining has agreed to contribute $9 million, towards the standard gauge railway leading from Kalgoorlie to Esperance in return for certain freight concessions from the State Government and the right to build and operate a private standard gauge railway to serve areas north of Kalgoorlie. Provision was also made in the agreement with the Government for Western Mining to retain the lateric nickel deposits at Ora Banda, which had indicated a likelihood of 120 million tons of 1 per cent nickel. Fortunately as the nickel finds are beginning to fall into the shape of a north-south line, these will probably be serviced by a single railway.

Western Mining is emerging as the leading nickel company of the State. By building its own smelter, it will be able to treat not only its own ore, but also the ore from other companies.

As we continued on our journey southwards, one of the passengers on the bus said we had reached the fourteenth mile near the place where Detective Inspector Walsh and Sergeant Pitman were murdered in 1926. This tragic drama occurred when the detectives found three men, named Evans, Coulter, and Treffene, treating illicit gold in the bush. The detectives were shot, and the men attempted to dispose of their bodies by dismembering and partly burning them in the furnace used to smelt the gold. Finally they dumped the remains down an abandoned shaft.

Later, the three men were arrested and charged with wilful murder. At the trial, Evans gave evidence against his associates and was pardoned, but Coulter and Treffene were hanged.

The bus sped along the road and next we passed the grey dumps and the leaning, wooden head-frame belonging to the shaft of the old Celebration Mine. This mine supported a small township in the 1930s, but all that is left to tell the tale now is a stony waste of ground. There were several other "shows" and mines working in this area. The Golden Hope and the White Hope produced good gold a few miles away. After the Golden Hope closed down, brackish water was pumped from its shaft and sent along a wooden pipe-line to help keep the plant working at the Celebration Mine.

This district faced the same problem that has blighted the entire Eastern Goldfields—namely, the lack of water. A little further on from the Celebration, we passed the Wollubar Dam. It was built in the middle of a large water catchment area and was originally a native soak. In 1903, water was pumped ten miles to serve the first Kambalda townsite by a steam driven pump, and the donkey boiler used to make the steam still stands near the dam. The water from Wollubar Dam was declared unfit for human consumption after the daughter of John Morgan contracted typhoid fever and spent a month in the Government Hospital in Kalgoorlie.

At the turn of the century a hotel opened close to the dam to cater for men prospecting in the district.

A brass plaque, to mark the spot near where surveyor C. C. Hunt blazed a kurrajong tree during an expedition in 1863, has been erected by the local branch of the Historical Society at White Hope. In 1919 a twelve foot lode worth one ounce of gold to the ton was discovered by White Hope Company near this tree.

Also in the area are the remains of the old King Battery erected about 1902-3; it was owned by the Hampton Plains Pastoral Company. It crushed ore, railed to it by a small train from two of its mines, the Merry Hampton and the Hampton Boulder, a few miles away. The remains of a thirty-foot wheel, used in the treatment of the sands left behind after crushing, can still be seen at the battery.

Our road wound its way through bush-covered hills and soon we were at Kambalda. After a short stay in the first nickel town, we drove on to Widgiemooltha, twenty-seven miles away. The Aboriginal meaning of this word is "Place of Evil Spirits," and very few natives will camp in the area. Now the tiny town of Widgiemooltha is the centre of a lot of activity. Western Mining transfers its containers of nickel concentrates, brought down by road from Kambalda, to the railway at Widgiemooltha siding. However, this is only a temporary measure until the new standard railway is built to Kambalda and so save the double handling. Salt mined from Lake Lefroy is also loaded at the siding to be taken to Esperance, then shipped overseas to Japan.

It is envisaged that Widgiemooltha will become a major nickel centre as work has begun on four mine shafts in the district. Some of the world's heavyweights in mining have done extensive exploration work in the Widgiemooltha area, but people outside those organisations can only guess at the result. The companies are renowned for their secrecy as well as for the vast amount of money they are spending on exploration.

A district landmark, two miles north of the town, is the big, yellow painted head-frame on the shaft being sunk by the International Nickel-Broken Hill Proprietary partnership. Two more shafts, one at Red Ross (named after Ross of the red beard who helped in the exploration) fifteen miles south of Widgiemooltha and the other at Wannaway, thirteen miles to the north-west, are being sunk in a joint venture by Cozinc Riotinto of Australia, New Broken Hill, and Anaconda. There is much speculation about the future of Widgiemooltha, but plans to build a $350,000 motel there have been approved in principle by the Coolgardie Shire Council and it is said that a joint township is planned by companies with interests in the area.

In latter years, the biggest mine in the Widgiemooltha area was the Paris Gold Mine. It was successfully worked by Mr Lister and his sons for over twenty years from 1931 until the Northern Minerals Syndicate took over control and began production in 1959. The mine finally closed down in 1964. The Paris Gold Mine was twenty-three miles from Widgiemooltha, and gold was originally found about 200 yards from the mine in a pothole by Mr French in 1920. This find caused much pegging, but no additional work was done in the area until 1925 when a Kalgoorlie syndicate, under the name of Paris Gold Mines, located the main reef. The total production from this centre was 60,155 tons yielding 23,966 fine ounces of gold and 17,094 fine ounces of silver. High values were also obtained from copper in the Paris Gold Mines.

At Widgiemooltha, a mine called the "Mount" produced 3,585 tons of ore for 912 fine ounces of gold and 486 fine ounces of silver till it closed in 1963. The town was the centre of a lot of prospecting activity and was also a welcome resting place for travellers between Esperance and Coolgardie at the turn of the century. In 1900, James Doyle ran the Widgiemooltha Hotel where the Coolgardie passenger coaches called regularly.

Some miners around the town were willing to experiment. One of their projects was reported in an article, "Novel Crushing Plant," in the *Norseman Pioneer* as follows: "An interesting trial run was given a novel battery at Widgiemooltha last week. The battery consists of 2 heads of stamps, and is driven by a man sitting on a bicycle, which is fixed in a solid frame, with rope around the hind wheel connecting with a pulley on a fly wheel shaft, which also has a pulley connecting by a rope with a cam-shaft pulley, thus having two indirect actions.

"The battery box is made out of a hollow tree, with a small 16 inch by 12 inch screen, whence the tailings pass over copper plates as in an ordinary battery. This novel affair is due to the enterprise of Messrs Hawkins and Rooke, who were hard pressed for crushing facilities to treat stone from a rich leader which they possess. The cams were made by a blacksmith, everything but the copper plates being the work of the prospectors. The trial run promised practical success."

Widgiemooltha had its share of eccentric characters and one was an Irishman who also had a novel idea. He kept pigs near the town and was reported to the R.S.P.C.A. for putting them out on the salt lakes, where they all perished. In his defence, he explained, "Well I figured it out this way: if you feed pigs ordinary food, you just get pork and I wanted salted pork so I put them out on the salt lake to graze."

For a few months in 1935, I lived with my family at Spargoville and attended school at Widgiemooltha for a week. The school was only a small hall and one teacher taught all classes. There were only about twenty children attending the school, but my young sister and I found it a welcome change from our correspondence lessons.

Spargoville is another name which will once again mean something in the mining world. In December 1970 Selcast Exploration announced that they had an economic nickel ore-body at Spargoville and they would sink a shaft.

My mother, sister and I went to Spargoville, nineteen miles north of Widgiemooltha, to join my father who was helping to develop the gold mine, Spargo's Reward, in 1935. It was then a small tent settlement and we were the only children except for Mrs Brown's young baby.

I will never forget our first night at Spargoville. We travelled by train from Perth to Coolgardie where my father met us. As we drove the twenty-seven miles to Spargoville in our blue Whippet utility, he told us, "I've built us a good camp near the mine. There's a big bough-shed where we can have our meals, and it also covers the two tents we'll be sleeping in. I haven't quite finished thatching the sides of the shed yet as I've been on night shift, but it will get done in a day or two."

We children thought our new home was lovely—better than living in an ordinary house—but I am not sure that our mother shared the opinion, especially when she discovered that she had to do her cooking on an outside fireplace and in a camp oven. Still, she bore up bravely.

That night, a sudden summer thunder storm blew up and the wind screamed through the tops of the gum trees all around our camp. At the height of the storm, a tall gum near the camp crashed down, resting partly on the roof of our utility and the rest crashing through the bough-shed, knocking the mainstay of our parents' tent down, and pinning them on the bed. The wind had lifted the rest of the partly thatched shed and had brought that down too. Fortunately only one side of our tent had collapsed, so my sister and I were unharmed.

Dad managed to crawl off the bed, telling Mum to get under it for safety. "I can't get off the thing to get under it" was her muffled reply, but she finally managed it.

Meanwhile, Dad crawled out under the debris of boughs and the clinging tent to see how my sister and I had fared. He crawled into the contents of the overturned Coolgardie cooler, and became covered in broken eggs and condensed milk. We spent the rest of the night in the only solid building in Spargoville at that time—a tin-roofed, hessian-walled house belonging to a married couple.

By the time the afternoon shift workers had come up from underground, the storm had cleared and the moon shone at intervals through scudding clouds. Walking past on their way home, the men looked over and saw the remains of our home and ran to see if we had been killed or were lying injured inside. One man crawled in under the main tent, and, seeing a round, white object, called out, "I can see poor old Harry's (my father) bald head." He groped under the bed only to find an enamel article in his hand.

The next morning, the men had a "busy bee" and built us a new home.

After we left the district, Spargo's Reward developed into a mine and real homes were built at Spargoville, but when the mine closed down at the end of the second World War, the town died.

There is only one house occupied at Spargoville now. It is owned by Mrs Eileen Turle, a woman prospector who has worked her own little "show" there for some time. She came out from England in 1949 with her husband, a poultry breeder, and their five children. Mr Turle was very interested in West Australian gold-mining and wanted to do field work and learn mining at first hand. In England, he had published brochures on mining in Western Australia, and had contributed to mining periodicals.

Armed with Sam Cash's *Loaming for Gold* the family went to Menzies, where Eileen Turle was to serve her apprenticeship as a prospector. They set up camp close to Menzies, near the stone cairn which Sir John Forrest had built on top of Mount Misery while surveying for the overland telegraph line. Mrs Turle tied old stockings to bushes to guide the children through the bush to the school at Menzies. The children had strict instructions not to leave one stocking before sighting another in case they became lost, an easy thing to do in that lonely bush country.

This versatile English family later moved to other locations, and while in Ravensthorpe they built their own home from old mining materials and bricks from an old smelter. They established a farm, and in their spare time they worked in an old copper mine where they had found an extension to a lode. By turning the handle of the windlass, Mrs Turle helped her husband "bring to grass" twenty tons of good grade copper ore. They also worked a small gold "show," the "Isabella," and extracted seven and a half tons of ore which paid for their next move, this time to the remains of the small gold-mining town of Spargoville. Here they pegged nearby Crown Land and did a lot of prospecting, particularly at Cave Rocks.

After the death of her husband, Mrs Turle continued working her "show" near the old Spargo's Reward Mine, as well as manning leases she held for her late husband's company. She often worked alone or sometimes with the help of a member of her family. The district around Spargoville is well known for its gem-stones and people go to Mrs Turle to obtain directions for fossicking, which she readily gives. Gem-stones are scattered on her desk and in odd corners of her cottage, but she has other interests besides mining. As well as being an artist and illustrator, she is a lover of nature. She says, "I am never lonely. Especially now the nickel boom is on. There is always someone camped close by, and I have a lot of visitors."

Now a mining company is interested in Mrs Turle's leases. If successful, this would be a just reward for her years of hard work and prospecting in the bush.

Our bus stayed in Widgiemooltha for twelve minutes, and then we left on the last stage of the journey to Norseman. As we travelled south, the bushes and gum trees became thicker. The sun glistened on the fresh gum tips, waving and tossing in the light wind. It was early Spring, and the country was almost at its best.

We passed Anaconda's big exploration camp, about thirteen miles south of Widgiemooltha. Then, except for a Main Road's Gang working on the road, there was little to break the monotony till we reached Norseman. The bus pulled in at the Norseman Railway Station a little ahead of time. The journey had taken under three hours, a lot quicker than the train trip which took the longer route via Coolgardie, and certainly faster than by horse or coach in the early days.

Top: The Scotia nickel mine as seen from the air. Bottom: An aerial view of the Golden Mile, looking towards Boulder

18
NORSEMAN

NORSEMAN IS STILL producing gold on the Central Norseman, although like the few other gold-mining areas left in the State it is struggling against the heavy overhead costs, labour shortages, and the need for a higher gold subsidy or a gold price rise. The other gold mine in the town, the Norseman Gold Mine, gave up gold mining in 1947 and concentrated on treating pyrites. There was a shortage of of sulphur, extracted from pyrites, since none was imported during the war years. Luckily a substantial body of massive pyrites was discovered during a drilling programme in 1939. The Iron King Mine, another source of pyrites, two miles from the town, was bought by Norseman Gold Mines for only one thousand dollars. After the Government contract ran out, the mine ceased producing pyrites and is now concentrating on mining salt at Lake Lefroy, near Widgie-mooltha.

Norseman has had its ups and downs, like other gold-mining centres. Now, like most goldfields towns, it is pinning its hopes for the future on nickel. During the days of gold its peak year was in 1905 when it had a population of

TOP: The Power Station at Kambalda, with Lake Lefroy in the background.
BOTTOM: A partial view of Kambalda East Township. Single men's quarters are in the foreground, and homes for married employees at the foot of the hill

5,000, and it sank to its lowest ebb in 1925 when there were only 300 people in the town. The present population is about 3,000.

The gold price rise and the discovery of the rich Butterfly leases by Norseman Gold Mines in the early 1930s revived the town. The slime dumps from this Butterfly reef are now a pale pink ochre colour in contrast to the plasticine grey of the larger dumps from the Central Norseman Mine close-by.

It was not until 1937 that the Goldfields Water Scheme piped water to Norseman. Till then the towns had to rely on unreliable and expensive water supplies: firstly desalination from the surrounding salt lakes, then water brought in tanks by train after the railway was built from Coolgardie to Esperance in the early 1930s. In spite of the shortage of water in those early years, the town boasted a brewery.

Norseman is in the district where the Dundas goldfield was proclaimed in August 1893, nearly a year after Bayley and Ford found gold at Coolgardie. A townsite laid out near Moganyer Soak, and the surrounding Dundas Hills were named after the Secretary of State for the Colonies at the time by Surveyor Roe, who was searching for pastoral land in the district in 1848. The first colours of alluvial gold were found by William Moir of Fanny's Cove when he was droving stock from Esperance. This inspired him to organise a prospecting party in 1892, but they failed to find additional gold.

At the same time, other prospecting groups were in the area. W. Mawson and R. Kirkpatrick discovered an auriferous reef on the west side of Lake Dundas, and called it the "May Bell." The men had to travel 200 miles through virgin bush to register their claims because the nearest registrar was at Southern Cross. However, almost immmediately after the first discovery, the Great Dundas and Scotia mines were announced and Warden Hicks took up residence at Esperance for the registering of further claims. A year later, on 13 August 1894, two applications were made for the Reward Claims fourteen miles to the north of Dundas. One was lodged by Sinclair and Allsop for their

134

discovery, the "Norseman," and the other by Ramsay, Talbot and Goodcliffe for their find, the "Mount Barker." These men are recognised as the "Five Founders of Norseman." The town was named after Sinclair's horse.

Other gold finds soom followed, including the very rich "Princess Royal," five miles north-east of Norseman. Captain Treloar was one of its first managers, and besides being a Justice of the Peace and a first-class cricketer, he was highly regarded as a mine manager.

In its early years, the Princess Royal gave consistent returns of high value and it was common knowledge that a big leakage of gold was taking place from the mine. There were rumours that parcels of gold were hidden somewhere around Trigg Hill, not far from the mine. Even many years after the mine's heyday some people still believed the story. But as one man wrote, "A search over such a big, granite boulder strewn area would be a task equal to looking for a man's hat in the Pacific ocean."

The new finds began a flow of population and business to Norseman from Dundas. Now Dundas is abandoned, but people still visit the old townsite to picnic at the famous Dundas Rocks nearby. These huge granite boulders seem like forlorn, abandoned playthings of giants. The bush nearby is an ideal spot for picnics, but unfortunately visitors have defaced the rocks by painting names and dates on them.

A community distinct from other goldfield towns sprang up in Norseman because most of the prospectors came from the coastal towns. The equipment needed for the mines was brought up from the coast too. Hundreds of tons of machinery were shipped into Esperance by gallant little vessels like the *Flinders*, *MacGregor*, *Helen Nichol*, *Storm Bird*, *Innaminka*, and *Rob Roy*. Then teamsters carted it through the harsh mallee country, and over the sandy soil of the sandplains to Norseman, 120 miles away. The men used camels, donkeys, and horses to pull the heavily laden wagons. Usually a team of fourteen or fifteen horses were used to draw a wagon carrying four tons. It took a team two or three weeks to make the journey, but after heavy rains it would be lucky to progress more than a mile or two a day. Often 100 teams carrying food and machinery were on the road at one time.

In winter, the teams had to contend with the vehicles becoming bogged, and in the summer with the terrific heat. Also there was little love lost between the teamsters and the coach drivers. They would have angry verbal exchanges over who should give way for the other to pass.

Norseman was proclaimed a Municipality on 17 January 1896 and a Roads Board was formed. Mr Arthur Austin was the town's first mayor. In 1894, a monthly mail service by packhorse was established between Esperance and Norseman. Later a weekly, then twice weekly, coach service took over. The mail service from Coolgardie was by bicycle. The hardy cyclist's journey through the bush in all weathers must have been anything but fun. After the railway went through to Coolgardie from Southern Cross in 1896, a bi-weekly coach service by Cobb and Co. replaced the bicycle mail.

During the "Roaring Nineties" Norseman had two news-papers, the *Miner* and the *Pioneer*, which were outspoken and full of fight. They were combined in 1898 to form the *Norseman Times*. In 1903, an opposition newspaper called *The Norseman Sentinel and Princess Royal Watchdog*, which was little more than a scandal sheet did not appeal to many of the townspeople. In one of its articles, it said that abortions were taking place in the town, and the following was printed in its gossip column: "That pretty little juice jerker at the 'Royal' has quite an aversion to wearing the orthodox feminine appendage, viz., corsets. The aforesaid little bit of lamb and salad is of the opinion that such unnecessary appendages are somewhat cumber-some when out for a constitutional stroll with her boy." This newspaper had a very short life.

"Mug Plumber" wrote of his memories of Norseman in the *Western Mail* in 1941. He described how Queen Victoria's Jubilee was celebrated in grand style. "There were a couple of horse races and foot races for the blacks, and by jove some of those gins could run. They showed a lot of leg when running, as their dresses were not the best. There was a greasy pig to catch, but the pig refused to run. After the sports were over, there was a big bon-fire lit at night in the main street and a hogshead of beer was rolled out and stuck in the fork of a tree. In less than an hour, there were drunken men lying all around the keg. Then the blacks were lined up and given a paper bag of tucker and a prize for the prettiest gin. There were none of them very handsome. I think they gave the prize to the ugliest."

As in most of the early goldfields towns, there was a lot of "make do" in some of Norseman's buildings. A landmark for many years was one of the earliest residences of the town called "Tin Dog House." It was made of empty "tinned dog" cans, opened and flattened out.

An old resident, Mr Scholloman, who at eighty-five still worked at his trade of a carpenter on one of the mines, reminisced: "The town has been a good place to live in. Very little trouble. We did have a beer strike once, and when some of the lads got dry they rolled a barrel of beer from the hotel to the railway station, and everyone had a drink. They paid for it after the strike was over."

Now Norseman is an important link between the West and the Eastern States, as it is the western beginning of the Eyre Highway. The local police force is kept busy, watching for stolen cars and wanted criminals using the Eyre Highway, and stranded hitch-hikers often pose problems. In 1970 three young travellers were bashed and robbed while asleep in their camp on the side of the Highway, and another man was murdered by a hitch-hiker not long afterwards. After these tragedies, drivers became wary of giving lifts to hitch-hikers, and many of those unable to hitch a ride were gaoled for vagrancy.

Perhaps Norseman is set in the prettiest surroundings of all the goldfields towns. From the top of the hill where the Homing Beacon sends out its beams to guide the planes safely on their way to and from the Eastern States, there is a glorious panorama of the country for miles around. The many hills are clothed with salmon gums, mulga, and several species of acacia which grow up to fifteen feet in height.

Lakes, which vary in colour from bright red ochre to pale clay, lazily wind their way around the town. The local airfield was built on one of the lakes and is hazardous in wet weather. There are deposits of gypsum close to the town and two enterprising brothers have used this to help make plaster-board locally. A lot of interest has been shown in a potential nickel area called the Jimbalana Dyke, a long ridge, or fault, west of the town. All the country for miles around Norseman—across the lakes, over the hills and through the bush—is pegged for nickel and associated minerals. Streamers of various colours flutter from the pegs.

Like other towns in the new nickelfields, rumours are rife. At a time when the share market was yo-yoing and very touchy, one exploration company's shares shot up suddenly. We found out later that the only reason for the price rise was that some of the field hands working for the company had been seen cutting down trees.

Nevertheless the townspeople have hopes that among all the leases pegged around their town, some will be mined for nickel and give employment to men when the town's last gold mine finally closes down. Then Norseman's future will be assured, and, like Kalgoorlie and Boulder After Nickel, it can feel secure once again.

19
AFTER NICKEL

THE TWIN TOWNS of Kalgoorlie and Boulder in 1972 A.N. (After Nickel) are exciting and vital places to live in. Now, there are voices of optimism everywhere. Although the gold mines had little life left, the new nickel mines of the Eastern Goldfields are coming to take their place.

But old habits die hard, so it will take us a little more time before we can call our part of the country the Eastern Nickelfields, as our ties and interest in gold-mining are still too strong.

Kalgoorlie and Boulder are now young folk's towns filled with mining men, exploration crews, drillers, draughtsmen, assayers, office personnel, and businessmen. The resemblance between these and the early gold-miners is mainly that of their beards, optimism, and enthusiasm.

Our children came back and found plenty of entertainment and night life. Now there are two night clubs, Chinese restaurants, steak bars, and good local and imported bands to play in the modernised hotels and television. There are also two new motels and a rival for the Palace Hotel in the beautiful, new Tower Motor Hotel. Kalgoorlie has become the place where the nickel men come to relax in their times off from work.

In two of Britain's newspapers, *News of the World* and *People*, Kalgoorlie was described early in 1970 as a "24 hour a day Sin Centre." *People* described the search for nickel as the greatest get-rich-quick gamble of the century and the *News of the World* said the boom was one of the greatest rushes for riches in history and that it was reminiscent of the exciting days in the 1880s. The paper conjured up a picture of Kalgoorlie as a rollicking twenty-four-hours-a-day sin centre where anything goes. There were "come-on" girls and card sharps operating under the noses of the police. It was a town "where girl-hungry, beer-thirsty men who spend months at a time in the sand-blown, fly-ridden out-back were only too happy to spend till their money ran out."

People said that besides its pubs, Kalgoorlie had baccarat and two-up schools, and, of course, the Hay Street girls, and that "a cruder form of prostitution can hardly be imagined."

The local reaction to these reports was varied. Some said that they were grossly exaggerated while others thought they could not be ignored. One person said that two-up, prostitution, and late hotel trading had been part of tradition throughout Kalgoorlie's history. "They are the only reason a number of miners stay here," he said. "As for 'come on' girls, I have never heard of them."

Another resident said that two-up had been tolerated, and was the "weekly Sunday outing" for many men. However, baccarat was relatively new as it had only been introduced in September 1969, when a school opened during the annual racing carnival.

Like every town, Kalgoorlie has many faces and the newspaper articles only showed our "bad" side. The gold-fields have always had a strong sporting community, with games and sport to suit every taste, and now Kambalda has added a team to local league football and introduced land yachting. When our drought breaks, and we get heavy summer rains to make the creeks run strongly, we have water sports a few miles from Kalgoorlie near Seven Mile Hill. There has already been some water-skiing on Lake Gidgie and other lakes after heavy rains, but our artificial lake, Lake Douglas, named after Doug Daws, one of the enthusiasts who helped get the scheme under way, will be the venue for all types of water sports.

The artificial lake and picnic spot was the brainchild of Mrs Margaret Bull, ex-headmistress of the Kalgoorlie Infants' School. In November 1970 a fountain was built at the junction of Wilson Street and Federal Road and named in her honour for her interest and help in local organisations and for her work over many years with the children of the town. Margaret Bull pursued her artificial lake project with vigour. She gathered around her a band of enthusiasts who helped by searching for a suitable site, raising money, and getting down to plain hard work. For the dream to become a reality, nature now has to play its part and send enough rain to fill the lake.

The towns are becoming more interested in culture and a Council was formed to foster the Arts. Through their efforts, we have been able to attend more plays, ballet, opera, recitals, and choir singing concerts, while art exhibitions and arts and craft displays are often presented. Soon we expect to have a branch of the State Library.

There have been some recent changes in Victoria Park. In 1970, a Home for the Frail Aged was opened in a portion of the park by the Little Sisters of the Poor. The proposal to build the Home there caused some controversy in the town. However, the park's main features were retained, including the Hoover Tree and the Rotunda where brass bands once entertained the many families and visitors on Sunday afternoons. It is fitting that the pioneers, to whom the park meant so much in their youth, should be able to enjoy it outside the doors of their Home in the evening of their lives.

In recent years Victoria Park has become less popular, perhaps mainly because most homes have their own lawns and the new Kingsbury Park is a more convenient picnic spot for those attending the Olympic Pool nearby.

Hannan Street is also showing signs of the times. Fortunately it has not yet lost its identity, but if the demolition and re-building that threaten are carried out it could well do so. The National Trust has placed the Public Buildings, including the Post Office, on its list of buildings in the State to be preserved. Some people have called for the preservation of the whole of Hannan Street. Failing that, perhaps a local awareness and pride of our unique town could be the best weapon to save it. Even if only the façades were preserved while the rest of the buildings are renovated or re-built, Hannan Street would retain the appearance of the main street of Kalgoorlie, the town that gold built.

Two hotels in the third block of Hannan Street, as well as the old market building across the road, are a *must* to be saved. They all have their own distinctive architecture. The York and the Oriental hotels, a few doors away from each other, are rather like decorated wedding cakes. When they were built, the owners must have been vying to outdo each other with their lavish *décor*. Further down Hannan Street is the Star and Garter Hotel, which had been severely damaged by a heavy diesel truck crashing through the front of the stone hotel. The façade was re-built in a modern style, so now the old hotel hides behind a new face. In the past few years it has become a tradition that, on Saint Patrick's Day, Mine Host entertains the Irish customers by wearing a bowler hat while serving green beer.

Dominating the sky-line at the top end of Hannan Street is the poppet-head of the Mount Charlotte Gold Mine which is still treating large quantities of low-grade ore. Especially when lit up at night, it gives a perfect back-drop for a mining town.

At least one section of Kalgoorlie was made aware of the mine's presence one day towards the end of 1970. Sixty tons of explosives were detonated underground and shifted 350,000 tons of ore on a 400 foot plane between the 500 feet and 600 feet levels. A few seconds after, the explosion dust began billowing out of several shafts on the outskirts of the town. It fused together and created a thick cloud of dust which descended on the town and stopped traffic for about two minutes. Red dust covered everything in range.

142

This was the third in a series of three pillar blasting operations at Mount Charlotte within a period of nearly two years. The first blast, in 1969, produced about 300,000 tons of ore, and the second 500,000 tons. Mount Charlotte, run by Gold Mines of Kalgoorlie Australia Ltd, commenced production in 1963, and is a unique mining operation. The mine is operated on the open-cut method so that the big ore-carrying trucks can drive down to collect the ore. Approximately three million tons of ore have already been mined.

Facing the mine, near the old Hannan Street Railway Station which was the beginning of the loop-line to Boulder, the mines, and back to Kalgoorlie, is our museum. It was once the British Arms Hotel, the smallest hotel in Western Australia, but it has since been renovated and now houses relics of Kalgoorlie's past.

Nearby a new two-storey block of brick and tile flats has been built. They are named the Wilson Flats in honour of Reg Wilson, a Town Councillor of many years' standing. Some other new buildings are dotted among the old, but so far the character of Hannan Street has not been altered too greatly.

An indication of the town's prosperity is the number of thriving businesses in shops that had previously been empty for years. There are drilling supply firms, exploration companies' offices, and many others. However, the real sign of the times is the fluttering, thickly packed mining notices on the board in front of the Mines Department.

Hannan Street now has traffic lights at its intersection, installed in 1969. There will be no more revelries around the King Pole in the middle of the intersection on future New Year Eves. The pole was taken down when the lights were installed. On New Year's Eve 1968, some among the crowd got out of control and there were some wild scenes in which a police sergeant was injured. The celebrations were more peaceful the next year when the main attraction was a man who climbed the King Pole to hoist a woman's stocking at the top.

There is a story of how the intersection was placed at the

corner of Maritana and Hannan Streets, instead of the logical choice of a block further down. Arthur Reid writes that: "While the town was being developed, the owner of the Commercial Hotel, on a corner of Cassidy and Hannan Streets, was approached by a surveyor. 'I've received instructions to survey a road from Hannans (Kalgoorlie) to the Great Boulder. By starting from this intersection, it will run in a straight line to the mine and will make this the main intersection and yours the principal hotel. It makes no difference to me where the road starts from and I can just as easily start it from Maritana Street, only there'll be a slight turn if it starts from there. But it will make a big difference to you. Is it worth £100 to start it from here?' 'No, it is not,' the owner replied. He later said that was the shortest and most expensive answer he had ever given.

The surveyor then approached the owner of one of the hotels on a corner of Maritana Street with the same proposition. Without any argument he was handed over 100 golden sovereigns. And that is the story of how Boulder Road had its beginnings from Maritana Street instead of Cassidy Street, the natural starting point."

Among the many new buildings in the district is the new Agricola College, opposite the Christian Brothers College in Wilson Street. Agricola College is a residential college for men and women attending the School of Mines, which is now a branch of the West Australian Institute of Technology. The college opened in 1971 and was the first of its kind in the Eastern Goldfields. The students will be some of the future metallurgists, engineers, geologists, and surveyors for the new nickel mines.

A new girls' Catholic high school, Prendiville College was opened in 1971 and plans are under way for the new Eastern Goldfields High School at South Kalgoorlie. There are two new native hostels in Boulder, one for the young Aborigines attending High School and another for native lads working in the goldfields area.

Another new building is the Boulder Shire Council's offices in South Kalgoorlie. The building, made of brick and tile, is decorated with a model poppet-head on one of

144

its front walls, and a spray of water can be seen spurting from an inner courtyard. It is a building which the town can be proud of.

Boulder almost lost its identity early in 1969 when more than fifty ratepayers in the Boulder Town Council area signed a petition seeking the abolition of the Council. In February, the Boundaries Commission took evidence at Boulder and recommended that the Boulder Council should be abolished and amalgamated with the Kalgoorlie Shire Council. The recommendation was accepted by the Minister, and, after Boulder candidates fought for equal representation on the new council, an election to select a council to administer the combined area was advertised. The abolition of the Boulder Town Council took place on 30 June 1969, and the Kalgoorlie Shire Council was dissolved. Although still on the map, Boulder lost its identity as a town after the Town Council was abolished, and for this reason the residents of Boulder turned out in full force to vote at the election in October. The result was a resounding victory for the Boulder candidates and only two of the former Kalgoorlie Shire members won back their seats. The new members thought it would be fitting to change the name of the Kalgoorlie Shire to the Boulder Shire Council, as Kalgoorlie was already represented in the Town of Kalgoorlie. The Minister for Local Government, Mr. Logan, approved the change and it became official in December 1969.

To house and cater for the rising population in the towns, the Minister for Housing, Mr O'Neill, announced in February 1970 a proposal to provide land for more than 800 residential units on a 300 acre site in the South Kalgoorlie area. When the scheme is undertaken, it will be the biggest single land development project in the history of the goldfields. The State Housing Commission will handle fifteen per cent of the construction of dwellings, while the rest will be undertaken by companies, private individuals, and developers wishing to build homes in the area. The project falls entirely within the boundaries of the Boulder Shire Council.

A lot of homes have already been built in South Kalgoorlie. Lake View and Star Ltd recently undertook a large building programme near the View Way Drive-in Theatre. The aim of the project was mainly to house migrants brought to the district to increase the mines' work force. Since then, other companies and individuals have built homes nearby. A Mexican style home is the show-place of the towns—and the envy of every housewife. Several new homes have been built between the old at Lamington, and in Killarney Street some of the new homes, complete with large swimming pools, would not be out of place in any fashionable suburb in the city.

The new brick buildings give a feeling of permanence to the towns and reflect the confidence in the future. There are already many buildings made of brick and stone; most homes were previously made of material which could have been easily pulled down and rebuilt if the mines had failed.

Since the discovery of nickel, millions of dollars have been spent by big commercial ventures and heavy expenditure is expected to continue for some time. The Government Hospital and the police station are among the public buildings being renovated and enlarged. A new courthouse, to be built opposite the police station in Kalgoorlie, is planned at a cost of $400,000.

Householders have added to the "new look" of Kalgoorlie and Boulder by renovating and painting their homes which are now valued at about three times their original value. Kalgoorlie has entered the Jet Age too, as we now have a jet air service. There are also plans to extend the airfield, or build a new one, and Kalgoorlie is being considered as an alternate International Airport.

Another new feature of the district is the marshalling yards built at West Kalgoorlie for the Standard Gauge Railway which now crosses the continent from east to west and cost two million dollars. Glamour came through Kalgoorlie in March 1970 when the all-Australian built, air-conditioned luxury express, the *Indian Pacific*, made its first public crossing of the continent on the new standard gauge railway. The 2,461 mile journey took about sixty-four hours, cutting nineteen and a half hours from the old Sydney-Perth passenger service.

Five months later, 130 steam train enthusiasts travelled in what could possibly be the last steam train to cross Australia from coast to coast. The passengers, mostly members of the New South Wales Transport Museum, paid $300 for the return trip. Several hundred local residents, including hundreds of school-children, turned out to greet the train, the *Western Endeavour*. It stopped for a while at the Kalgoorlie railway station before continuing its journey to Perth.

Visitors of all races, nationalities, and ranks have visited the goldfields since the discovery of nickel. Among them were some of the geologists attending an International Symposium on Archaean Rocks held in Perth in May 1970. It attracted 614 geologists from all parts of Australia, as well as from overseas, and was the biggest gathering of geologists ever to be held in the Southern Hemisphere.

The mining areas of Western Australia are located on one of the largest, partially explored Archaean regions in the world. The area is roughly a triangle—Kalgoorlie to Pemberton to Port Hedland—and includes the old mining towns plus the new mineral regions. Naturally the area was of great interest to the geologists.

The saddest aspect of the new nickelfields is the phasing out of the gold-mining industry, the industry that not only gave life to this part of the country but also brought prosperity to the State and trebled its population in a short time after the discovery of gold. The gold-mining industry could have had a sudden death in June 1970 if the Federal Government had not extended the gold subsidy of eight dollars an ounce for a further three years. The mines were struggling along with a reduced work force and against rising wages and costs. A subsidy of twelve dollars an ounce had been asked for to help them keep going for a few more years until the new nickel mines came into production, but the increase was denied. Mr Hartrey, the solicitor who has been closely associated with the movement to increase the subsidy under the Gold Mining Assistance Act, expressed the sad feelings of many people when he said, "The gold-fields have been given three years in which to die and be interred."

Other requests for the increased subsidy were later forwarded to Canberra, but a petition, signed by goldfields residents, was refused in January 1971. The Finance Editor of the *West Australian* wrote that "the Federal Government is obviously convinced that the gold-mining industry in this State is nearing extinction and was prepared to do nothing to prolong its life."

Sadly, it appears that the famous Golden Mile will soon be preparing for its final closure. Indeed, by 1973 Australia may be in the unusual position of becoming a net importer rather than an exporter of gold. If this does happen, it will be a sad day for the old goldminers. Even if the price of gold rises after the mines close, it is unlikely that they will be able to re-open again.

However, life goes on and, inevitably, changes, so we must look forward with confidence and hope to our future with nickel. But we must never forget our romantic past and the gold that built our heritage.